ÍNDICE

Nota del autor

This material contains enough theoretical development and proposed exercises with solutions, in order to promote our students' autonomy. Moreover, for almost each introduced concept, some exercises are solved step by step.

To prepare this material, I have used the following bibliography:

Bibliography	- Fundamentals of Mathematics (Authors: J. Van Dyke, J. Rogers, H. Adams; Ed.: BROOKS/COLE, CENGAGE Learning),
	- Spanish text books:
	   - Matemáticas 1º ESO (Autores: J. Cólera y otros; Ed.: Anaya).
	   - Matemáticas 1º ESO (Autores: A. M. Gaztelu y A. González; Ed.: Santillana).
	- Notes and work sheets (in English and Spanish): Alfonso González, José A. Jiménez Nieto, Miguel Ángel Hernández Lorenzo, Isabel Torres (IES Ruradia), IES Don Bosco.
	- English Dictionary (Oxford Edition).

This materials have been reviewed in order to fulfil new Spanish **LOMCE** regulations.

Para el profesor: <u>Evaluación por estándares:</u> He preparado un cuadro en el que relaciono cada estándar evaluable de la LOMCE con los ejercicios de este libro que lo trabajan, y que puede ser muy útil a la hora de diseñar las pruebas de evaluación. Si estás interesado en que te lo envíe, ponte en contacto conmigo en el mail de más abajo.

En mi web (www.cuadernodepitagoras.com) he subido las presentaciones que uso con mis alumnos en clase. Si en algo crees que te puedo ser de utilidad, si encuentras erratas, o si deseas hacerme alguna sugerencia, te agradeceré que te pongas en contacto conmigo en el mail javiersanchezpi@gmail.com.

Soy doctor en Química y licenciado en Ciencias Ambientales y soy profesor de matemáticas en un instituto en Murcia (España).

Espero te sea de utilidad y que contactes conmigo para lo que precises.

Un saludo.

Javier Sánchez Pina.

Unit 1.- Natural numbers

1. Remember how to read numbers

First of all, we are going to remember how numbers are read in English. If necessary, look at your notes of last course. Anyway, we are going to remember now. At the beginning of some units, we will also remember whatever we need to know about reading.

Ordinals and cardinals

Remember natural numbers are classified, depending on their utility, into:

– Counting: "There are three apples on the table" → CARDINAL NUMBERS

– Ordering: "Barcelona is the second largest city in our country" → ORDINAL NUMBERS

Cardinal Numbers	Ordinal Numbers	Cardinal Numbers	Ordinal Numbers
1. one	1st. First	20. twenty	20th. Twentieth
2. two	2nd. Second	21. twenty-one	21st. Twenty-first
3. three	3rd. Third	22. twenty-two	22nd. Twenty-second
4. four	4th. Fourth	23. twenty-three	23rd. Twenty-third
5. five	5th. Fifth	24. twenty-four	24th. Twenty-fourth
6. six	6th. Sixth	30. thirty	30th. Thirtieth
7. seven	7th. Seventh	40. forty	40th. Fortieth
8. eight	8th. Eighth	50. fifty	50th. Fiftieth
9. nine	9th. Ninth	60. sixty	60th. Sixtieth
10. ten	10th. Tenth	70. seventy	70th. Seventieth
11. eleven	11th. Eleventh	80. eighty	80th. Eigthtieth
12. twelve	12th. Twelfth	90. ninety	90th. Ninetieth
13. thirteen	13th. Thirteenth	100. a/one hundred	100th. Hundreth
14. fourteen	14th. Fourteenth	101. a/one hundred	101st. Hundred and first
15. fifteen	15th. Fifteenth	200. two hundred	200th. Two hundreth
16. sixteen	16th. Sixteenth	1 000. a/one thousand	1 000th. One thousandth
17. seventeen	17th. Seventeenth	10 000. ten thousand	10 000th. Ten thousandth
18. eighteen	18th. Eighteenth	100 000. a/one hundred thousand	100 000. One hundred thousandth
19. nineteen	19th. Nineteenth	1 000 000. a/one million	1 000 000 One millionth

Beyond a million, the names of the numbers differ depending where you live. The places are grouped by thousands in America and France, by the millions in Great Britain, Germany and Spain.

Name	American-French	English-German-Spanish
million	1,000,000	1,000,000
billion	1,000,000,000 (a thousand millions)	1,000,000,000,000 (a million millions)
trillion	1 with 12 zeros	1 with 18 zeros
quadrillion	1 with 15 zeros	1 with 24 zeros

AND

AND is used before the last to figures (tens and units) of a number.

> 103: a (or one) hundred **and** three
>
> 325: three hundred **and** twenty-five.
>
> 2315: two thousand three hundred **and** fifteen

A or ONE

The words hundred, thousand and million can be used in the singular form with "a" or "one", but not alone.

"A" is more common in an informal style; "one" is used when we were speaking more precisely.

> I want to live for **a** hundred years
>
> The journey took exactly **one** hundred light years
>
> I have **a** thousand euros

"A" is also common in an informal style with measurement-words.

> A kilo of oranges costs **a** pound
>
> Mix **one** liter of milk with **one** kilo of flour ...

SINGULAR OR PLURAL?

Number are usually written in singular.

> Three hundred euros
>
> Several thousand light years

The plural is only used with dozen, hundred, thousand, million, billion, if they are not modified by another number or expression (for example a few/several).

> Hundreds of pounds
>
> Thousands of light years

PHONE NUMBERS

Each digit is said separately (25 → two five)

The figure '0' is called oh (304 → three **oh** four)

Pause after groups of 3 or 4 figures

> 234 7809 → two three four, seven eight oh nine
>
> 234 5778 → two three four, five double seven eight (British English) or two three four, five seven seven eight (American English)

ZERO, NOUGHT, NIL, LOVE …

The figure 0 is usually called **nought** in British English and **zero** in American English

In measurements, 0 is called zero:

> Water freezes at zero degrees Celsius

In team games, zero scores are usually called nil in British English, and zero in American English. In tennis, the word love is used instead of zero (this is derived from French word "l'oeuf", because zero can be egg-shaped):

> Spain three Germany nil (zero)
>
> Nadal is winning forty-love

1 BILLION

In English, billion usually means a thousand million: 1 000 000 000

Realize that in Spanish billion means 1 000 000 000 000

PLACE VALUE

Every digit in a number represents a different value depending on its position.

For example: In 53, "5" represents fifty units

                     In 5234, "5" represents five thousand units

This is the place value table we need to write numbers, no matter how big they are:

hundred ten one	hundred ten one	hundred ten one	hundred ten one
Billion	milllion	thousand	(unit)

CALCULATIONS

ADDITION $2 + 4 = 6$	• Two **and/plus** four **is/are/equals** six • Two **added to** four **makes** six • **What's** two **and** four? **It's** six.
SUBTRACTION $8 - 5 = 3$	• Eight **minus** five **is/are/equals** three • Eight **take away** five is three • Five **from** eight **leaves/is** three
MULTIPLICATION $6 \cdot 5 = 30$	• Six **times** five **is/equals** thirty • **Six fives are** thirty • Six **multiplied by** five **is/makes** thirty (More formal way)

DIVISION 12 : 3 = 4	• Twelve **divided by** three **is/are/equals** four • Three **into** twelve **goes** four times (for smaller calculations)

POWERS 6^5 (6 is the base and 5 is the exponent)	• Six **to the power** of five • Six to **the fifth power** • Six **raised to** fifth	SPECIAL POWERS 5^2 : Five squared 4^3 : Four cubed

ROOTS $\sqrt{16} = 4$	• The **square root** of sixteen **is/equals** four.

Exercises

1. Write the following numbers:
  a) 3456   b) 90304   c) 765   d) 237   e) 98053
  f) 134008   g) 45004   h) 150003

2. Write the following numbers in words:
  a) 3528   b) 86 424   c) 987   d) 3270   e) 30001
  f) 1487070   g) 320569   h) 20890300

3. Write the missing words. Then, write the answers in numbers and symbols:
  Example: Ten plus three equals *thirteen 10+3=13* _____ .
  a) Twelve minus six equals _____
  b) Seven times one equals _____
  c) Twenty-five divided by five equals _____
  d) Eight plus four minus nine equals _____

4. Write the missing numbers. Then, write the answers in words.
  Example: 3+8 = 11. *Three plus eight equals eleven.*
  a) 3 · ___ = 30   _____
  b) ___ -5 = 13   _____
  c) ___ +2 = 4   _____
  d) ___ :2 = 10   _____

5. Write the missing symbols. Then, write the answers in words.

3 ___ 7 ___ 4 = 14 _____

9 ___ 2 ___ 2 = 20 _____

25 ___ 5 ___ 4 = 9 _____

6 ___ 3 ___ 2 = 16 _____

16 ___ 4 ___ 6 = 10 _____

7 ___ 2 ___ 4 = 9 _____

6. Calculate the following powers mentally and write them in words:

Example: $4^3 = 64$ *Four cubed equals sixty-four*

a) $5^4 =$ _____

b) $11^2 =$ _____

c) $2^5 =$ _____

d) $5^3 =$ _____

e) $10^3 =$ _____

f) $100^2 =$ _____

7. Calculate mentally and write in words as in the example:

Example: $\sqrt{16} = 4$ *The square root of sixteen is four*

a) $\sqrt{25} =$ _____

b) $\sqrt{121} =$ _____

c) $\sqrt{900} =$ _____

2. Writing names from numbers and vice versa

Writing names from numbers

Numbers larger than 9 are written in place value name by writing the digits in positions having standard place value. Word names are written words that represent numerals. The word name of 213 is two hundred thirteen.

In our written whole number system (called the Hindu-Arabic system), digits and commas are the only symbols used. This system is a positional base 10 (decimal) system.

The location of the digit determines its value, from right to left. The first three place value names are one, ten, and hundred. See Figure.

hundred	ten	one

Example 1: Write number 764 in its expanded form:

Solution:

7	6	4
HUNDRED	TEN	ONE

- 4 contributes 4 ones, or 4, to the value of the number.
- 6 contributes 6 tens, or 60, to the value of the number.
- 7 contributes 7 hundreds, or 700, to the value of the number.

So 764 is 7 hundreds + 6 tens + 4 ones, or 700 + 60 + 4.

These are called **expanded forms** of the number. The word name is seven hundred sixty-four.

For numbers bigger than 999, we use the following table:

hundred ten one	hundred ten one	hundred ten one	hundred ten one
Billion	milllion	thousand	(unit)

Be careful!!!

- In Spanish, 1 billion is equal to 1 million millions, but in English, 1 billion is 1000 million.
- When writing numbers, digits (1, 1, 2, ... , 9) are separated into blocks of three digits, by *commas*.
- Notice we do not say "two thousands or millions" but "two thousand or million".
- When reading a number, the word AND is read only before last two digits.

Example 2: Write number 63,506,345,222 in its expanded form:

Solution: We write it on the table:

63	506	345	222
billion	million	thousand	unit

The number is read "63 billion, 506 million, 345 thousand, 222." The word *units* for the units group is not read. The complete word name is sixty-three billion, five hundred six million, three hundred forty-five thousand, two hundred twenty-two.

Writing numbers from their names

To write the place value name from the word name of a number, we reverse the previous process. First identify the group names and then write each group name in the place value name. Remember to write a 0 for each missing place value.

Example 3: Write with digits: three billion, two hundred thirty-five million, nine thousand, four hundred thirteen.

Solution:

    Step 1: three <u>billion</u>, two hundred thirty-five <u>million</u>, nine thousand, four hundred thirteen.

    Step 2:   3 billion,   235 million,   9 thousand,   413 units.

    Step 3:   Now, we put them in their blocks, separately, with commas:

             3,235,009,413.

Exercises

8. Write the word names of each of these numbers.

    a) 843   b) 196   c) 460   d) 710   e) 7020   f) 66,086

9. Write the place value name.

    a) Eighty-seven   b) Thirty-nine   c) Nine thousand, five hundred

    d) Nine thousand, five   e) One hundred one million

10. Write the word name of each of these numbers.

a) 27,680   b) 27,068   c) 207,690   d) 270,069

e) 54,000,000   f) 780,000

11. Write the place value name.

a) Two hundred forty-three thousand

b) Three hundred fifty-nine thousand, seven hundred eight

c) Twenty-two thousand, five hundred seventy

d) Twenty-three thousand, four hundred seventy-seven

e) Nineteen billion

f) Nine hundred thousand, five

12. Write in words the following numbers as in the previous examples:

	In English:	In Spanish:
2,538		
762		
90,304		
8,300,690,285		
593		
1,237,569		
3,442,567,321		
76,421		
8,321,678		
250,005		

13. Write these numbers as digits.

a) Five thousand, three hundred and four =

b) Three thousand, five hundred and four =

c) Four thousand and five =

d) 5 thousands + 2 hundreds + 3 tens + 4 units =

e) 4 thousands + 7 tens + 2 units =

f) 23 units + 50 hundreds =

g) 3 hundreds + 52 tens + 6 units =

h) 5 thousands + 2 hundreds + 410 units =

14. Read the following numbers:

	In English:	In Spanish:
120,000.321		
453,897		
763,123		
8,300,345		

3. Order in natural numbers

Symbols $<$, $\leq$, $>$ and $\geq$ are used to show the size of one number compares to another.

$a < b$ a is **less** than b	$a \leq b$ a is **less or equal** than b
$a > b$ a is **greater than** b	$a \geq b$ a is **greater or equal** than b

15. Place the following numbers in order:

a) 14, 21, 9 :   __ < __ < __ (in ascending order)

b) 12, 40, 32:   __ > __ > __ (in descending order)

16. Complete, using "<", ">" or "=":

   a) 2___3   b) 4___7   c) 3___3   d) 6___1   e) 8___2   f) 4___4

17. Write these numbers in ascending order: $1 - 5 - 7 - 2 - 3 - 4$

18. Find the missing number:

   a) $4 < $ __ $ < 6$ b) $8 > $ __ $ > 6$ c) $2 < $ __ $ < 4$ d) $7 > $ __ $ > 5$

19. Write three more numbers that follow the same patterns.

   a) 3, 6, 9, 12, 15, …   b) 4, 7, 10, 13, 16, …   c) 15, 13, 11, …

20. Put these numbers in ascending (smallest to biggest) order:

   a) 23 117 5 374 13 89 67 54 716 18

   b) 1272 231 817 376 233 46 2319 494 73 1101

21. Write down the value of the number 4 in each of these:

   For example 408: four hundreds

   a) 347_____   e) 4897_____

   b) 6754_____   f) 6045_____

   c) 64098_____   g) 745320_____

   d) 41_____   h) 405759_____

22. Insert $<$ or $>$ between the numbers to make a true statement.

   a) 12   22   b) 53   49   c) 61   54   d) 62   71

   e) 246   251   f) 212   208   g) 7470   7850   h) 2751   2693

23. Join up each number to the corresponding point on the number line.

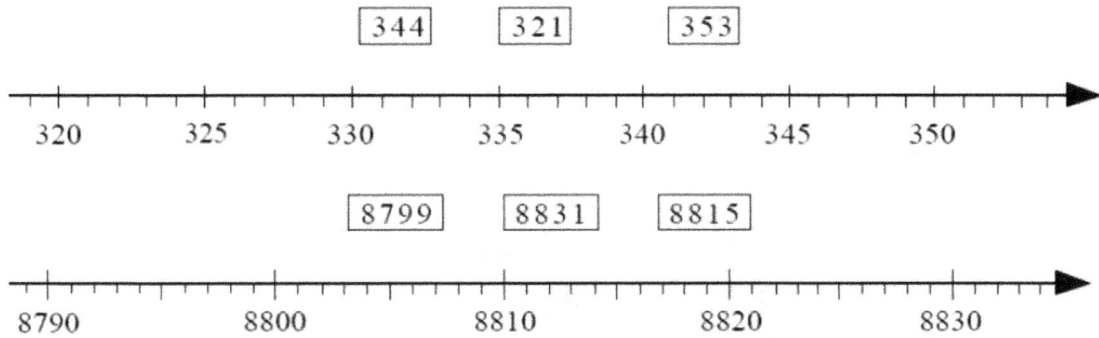

| 344 | | 321 | | 353 |

```
    320      325      330      335      340      345      350
```

| 8799 | | 8831 | | 8815 |

```
    8790        8800        8810        8820        8830
```

4. Roman numerals

I=1	(I with a bar is not used)
V=5	$\overline{V}$=5,000
X=10	$\overline{X}$=10,000
L=50	$\overline{L}$=50,000
C=100	$\overline{C}$ = 100 000
D=500	$\overline{D}$=500,000
M=1,000	$\overline{M}$=1,000,000

1 = I	11 = XI	25 = XXV
2 = II	12 = XII	30 = XXX
3 = III	13 = XIII	40 = XL
4 = IV	14 = XIV	49 = XLIX
5 = V	15 = XV	50 = L
6 = VI	16 = XVI	51 = LI
7 = VII	17 = XVII	60 = LX
8 = VIII	18 = XVIII	70 = LXX
9 = IX	19 = XIX	80 = LXXX
10 = X	20 = XX	90 = XC
	21 = XXI	99 = XCIX

- There is no zero in the Roman numeral system.

- The numbers are built starting from the largest number on the left, and adding smaller numbers to the right. All the numerals are then added together.

- The exception is the subtracted numerals, if a numeral is before a larger numeral; you subtract the first numeral from the second. That is, IX is 10 - 1= 9.

- This only works for **one** small numeral before one larger numeral - for example, IIX is not 8; it is not a recognized roman numeral.

- There is no place value in this system - the number III is 3, not 111.

Exercises

24. Translate these roman numbers into decimal system:

  a) XIV   b) LXXIII   c) LXIX   d) CCXVII   e) DCXC   f) MCMLVI

25. Write in romans:

  a) 18   b) 36   c) 54   d) 333   e) 608   f) 2 390

26. Write in romans:

  a) 630   b) 638   c) 639   d) 640   e) 2425   f) 2525

  g) 3001   h) 3520

Sol.: a) DCXXX; b) DCXXXVIII; c) DCXXXIX; d) DCXL;

e) MMCDXXV; f) MMDXXV; g) MMMI; h) MMMDXX.

---

5. Calculations

5.1. Sum (or addition)   + PLUS

In small additions we say *and* for + and *is/are* for =

  2+6 = 8   two **and** six **are** eight

  What's two **and** six? **It's** eight

In larger additions (and in more formal style) we use *plus* for +, and *equals* or *is* for =

  700 + 200 = 900 Seven hundred **plus** two hundred **equals** / **is** nine hundreds.

5.2. Subtraction   − MINUS

In a formal style, or with larger numbers, we use *minus* and *equals*

500 - 300 = 200 Five hundred **minus** three hundred **equals** /**is** two hundred

---

27. With the road information, calculate the following distances:

a) A CORUÑA – S. SEBASTIÁN

b) A CORUÑA – GIJÓN

c) GIJÓN - BILBAO

d) BILBAO – S. SEBASTIÁN

SANTANDER

A CORUÑA 448 km

BILBAO 107 km

GIJÓN 204 km

SAN SEBASTIÁN 228 km

A CORUÑA   GIJÓN   BILBAO   SAN SEBASTIÁN

28. Work out: a) $250 + 75 + 130$ b) $524 – 215 – 132$ c) $420 + 175 – 368$

29. Associate each of the following statements (1 to 3) with below expressions (a to f):

1) Rosa has got 13 € and buys a 8 € book, but they apply a 3 € discount.

2) Andrés has got 13 € and buys a 8 € commix and a 3 € notebook.

3) Marta had got 13 €, she has been given 8 € and she has given 3 € back to her sister.

a) $13 – 8 – 3$ b) $13 – 8 + 3$ c) $13 - (8 + 3)$

d) $13 - (8 - 3)$ e) $13 + (8 - 3)$ f) $13 + 8 – 3$

30. Calculate:

a) $52 - (25 - 13)$ b) $40 - (32 - 16)$ c) $28 + (11 - 6)$ d) $37 + (15 - 12)$

31. Scafell Pike is 979 m high and Ben Nevis is 1344 m high. What is the difference in height between the two mountains?

32. Robert's family are going to visit Seville this weekend. They expect to expend €100 on the hotel, €50 on a meal and €45 on petrol. At the moment they have € 200. Do they have enough money for their trip? What is the difference between the money they have and the money they need?

5.3. Product (or multiplication) TIMES MULTIPLIED BY

In small calculations we say

3 x 4 = 12 three fours **are** twelve

6 x 7 = 42 six sevens **are** forty-two

In larger calculations we can say

17 x 200 = 3400   17 **times** 200 **is/makes** 3400,

                or in a more formal style 17 **multiplied by** 200 **equals** 3400

Exercises

33. Multiply the following and read your calculations:

   a) $75 \cdot 5$ b) $847 \cdot 53$ c) $93 \cdot 65$ d) $413 \cdot 26$ e) $309 \cdot 61$

34. Calculate:

   a) $14 \cdot 10$ b) $14 \cdot 100$ c) $14 \cdot 1000$ d) $333 \cdot 10$ e) $333 \cdot 100$

   f) $333 \cdot 1000$ g) $5 \cdot 10$ h) $5 \cdot 100$ i) $5 \cdot 1000$

35. Calculate and compare the results:

   a) $5 \cdot (3 + 2)$ a') $5 \cdot 3 + 5 \cdot 2$ b) $4 \cdot (10 + 9)$ b') $4 \cdot 10 + 4 \cdot 9$

   c) $10 \cdot (11 - 1)$ c') $10 \cdot 11 - 10 \cdot 1$ d) $21 \cdot (4 - 2)$ d') $21 \cdot 4 - 1 \cdot 2$

Name the property that explains these results.

36. Calculate by using the distributive property and doing the calculations in parenthesis first. Do you get the same results?

   a) $5 \cdot (2 + 6)$ b) $10 \cdot (12 - 2)$ c) $11 \cdot (11 - 1)$ d) $26 \cdot (9 + 1)$

37. Work out the total cost of 6 pens at 56p each.

38. Calculate:

a) The number of seconds in one hour.

b) The number of seconds in a day.

c) The numbers of seconds in the month of October.

39. Paul has 16 heartbeats for 15 seconds. Calculate his number of heartbeats for one minute.

Determine the number of heartbeats for one hour, a day and a year.

40. Find the area of the following shape:

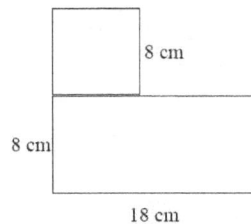

41. Calculate:

a) $347 \cdot 20$ b) $86 \cdot 50$ c) $1005 \cdot 280$ d) $41 \cdot 2500$ e) $32 \cdot 1516$ f) $99 \cdot 99$

42. Calculate without paper or pencil:

a) $3 \cdot (2 \cdot 5) \cdot 8$ b) $5 \cdot 7 \cdot 2 \cdot 4$ c) $6 \cdot 40$ d) $35 \cdot 8$

43. A delivery company truck is driven every Monday, Tuesday and Friday Lugo – Pontevedra (round trip). If distance between Lugo and Pontevedra is 148 km, how many kilometers are driven each week?

44. Check that each expression to the left is equivalent to its corresponding to the right.

a) $6 \cdot (3 + 5) \leftrightarrow 6 \cdot 3 + 6 \cdot 5$

b) $5 \cdot 9 - 5 \cdot 7 \leftrightarrow 5 \cdot (9 - 7)$

c) $10 \cdot 8 - 10 \cdot 6 \leftrightarrow 10 \cdot 2$

d) $8 \cdot 5 \leftrightarrow 8 \cdot 2 + 8 \cdot 3$

45. Calculate:

   a) 14·100   b) 82·1000   c) 1001·10   d) 52·10000   e) 80·100   f) 13000·10

5.4. Quotient (or division)   DIVIDED BY

Divisions are read as follows:

   270:3 = 90 Two hundred and seventy **divided by** three **equals** ninety

   But in smaller calculations (8:2 = 4) we can say two **into** eight **goes** four (times)

46. Do these questions:

   a) 96 : 6 =   b) 990 :   = 54   c) 834 :   = 18

   d) 747 :   = 9   e) 9165 : 37 =   f) 9564 : 74 =

47. Seven people share a lottery win of £868.00. How much does each person get?

In a division of natural numbers, we find four elements: D=dividend, d=divisor, q=quotient and r=remainder.

$$\begin{array}{c|c} D & d \\ \hline r & q \end{array}$$

For example, in the following division:

$$\begin{array}{c|c} 15 & 2 \\ \hline 1 & 7 \end{array}$$

The dividend is 15, the divisor is 2, the quotient is 7 and the remainder is 1. This division isn't **exact** (its remainder is not 0).

18

But in this division:

$$15 \quad | \underline{3 \qquad}$$
$$\;\, 0 \quad\;\; 5$$

the remainder is zero, so we say this division is **exact**.

If we want to know the division is correct, we can use the formula: $D = d \cdot q + r$

In the examples: $15 = 2 \cdot 7 + 1$ and $15 = 3 \cdot 5 + 0$.

<table>
<tr><td rowspan="1" style="writing-mode:vertical">Exercises</td><td>

48. Work out these calculations:

  a) 45:5   b) 235:5   c) 1476:12   d) 5175:15   e) 9913:23

49. Calculate:

  a) 250:10   b) 53000:10   c) 15000:100   d) 1000:10

50. Complete the following operations:

  a) $123 \cdot \boxed{\ldots} = 5\ 904$ b) $\boxed{\ldots} \cdot 86 = 1\ 548$ c) $\boxed{\ldots} : 57 = 26$ d) $1\ 862 : \boxed{\ldots} = 133$

51. In a division, divisor is 7, quotient is 13 and remainder is 5. What is the dividend?

52. Calculate the quotient and the remainder:

  a) 258 : 23   b) 14 315 : 47

53. How many 12 caramel bags can be completed with 250 caramels?

54. Eggs are packed in boxes of twelve. How many boxes do you need to pack 1476 eggs?

55. Lucas and his two friends earn 65 € working as waiters. Can the money be shared out exactly between them?

56. DVD's are packed in boxes of 25. How many boxes are needed to pack 1 028 DVD's? How many DVD's will be left over?

57. Susan drinks 21 liters of water every week. How much does she drink each day?

58. Richard buys 150 text messages for €5. What is the price of one text messages?

</td></tr>
</table>

59. Write the missing words. Write the answers in words

  a) Twelve minus seven equals _____

  b) Six times five equals _____

  c) Eighty minus seventeen is _____

  d) Forty four minus nine plus twenty three equals_____

  e) Three times fifteen divided by five equals_____

60. Write the missing numbers and write the answers in words as in the example

  Example: $3 + 14 = 17$ three plus fourteen equals seventeen

  a) $6 \times \underline{} = 42$ b) $18 - \underline{} = 11$ c) $60 : \underline{} = 2$ d) $12 + \underline{} = 25$

Parentheses ( ) and **brackets [ ]** are used in mathematics as **grouping** symbols. These symbols indicate that the operations inside are to be performed first. Other grouping symbols that are often used are **braces { }** and the **fraction bar** —.

Without a rule it is possible to interpret $6 + 4 \cdot 11$ in two ways:

$$6 + 4 \cdot 11 = \qquad\qquad \text{or} \qquad\qquad 6 + 4 \cdot 11 =$$
$$= 6 + \quad 44 \;=\; \boxed{50} \qquad\qquad\qquad = \quad 10 \;\cdot 11 = \boxed{110}$$

In order to decide which answer to use, we agree to use a standard set of rules. Among these is the rule that we multiply before adding. So, $6 + 4 \cdot 11 = 50$.

The order in which the operations are performed is important because the order often determines the answer. Therefore, there is an established *order of operations*. This established order was agreed upon many years ago and is programmed into most of today's calculators and computers.

5.5. Order of operations

1ST: Parentheses	**2ND**: Exponents
3RD: Multiplication/Division	**4TH**: Addition/Subtraction

Consider the phrase **Please Excuse My Dear Aunt Sally**. Note that the first letters of the words in this phrase are exactly the same (and in the same order) as the first letters for the order of operations.

Many students use "Please excuse my dear Aunt Sally" to help them remember the order for operations. Why not give it a try?

Exercises

61. Calculate:

a) $8 + 7 - 3 \cdot 4$ b) $15 - 2 \cdot 3 - 5$ c) $22 - 6 \cdot 3 + 5$

d) $36 - 8 \cdot 4 - 1$ e) $4 \cdot 7 - 13 - 2 \cdot 6$ f) $5 \cdot 4 + 12 - 6 \cdot 4$

g) $5 \cdot 6 - 4 \cdot 7 + 2 \cdot 5$ h) $8 \cdot 8 - 4 \cdot 6 - 5 \cdot 8$

62. Work out:

a) $2 \cdot (4 + 6)$ b) $2 \cdot 4 + 6$ c) $8 : (7 - 5)$ d) $5 \cdot 7 - 5$

e) $(5 + 6) \cdot 4$ f) $5 + 6 : 3$ g) $(19 - 7) : 2$ h) $18 - 7 \cdot 2$

63. Calulate:

a) $48 \div (3 + 5)$ b) $(5 + 4) \times 14$ c) $(40 + 30) \div 5$

d) $(27 + 21) \div 3$ e) $(22 + 33) \div 11$ f) $(40 \div 20) \times 3$

64. Calculate:

a) $3 + 6 \times 2 + 5$ b) $(4 + 3) \times 5 - 2$ c) $15 - 6 : 2 \times 4$

d) $15 - 16 : (3 + 1)$ e) $3 + 6 \times 2 + 10$ f) $(58 - 18) \times (27 + 13)$

g) $(32 - 8) : (6 - 3)$ h) $(32 - 8) : 6 - 3$ i) $67 + 16 \times 3$

65. Insert brackets to make the following calculations correct

a) $5 + 4 \, x \, 8 = 37$ b) $5 + 4 \, x \, 8 = 72$ c) $6 + 15 \div 3 = 11$

d) $6 + 15 : 3 = 7$ e) $5 + 4 + 3 \, x \, 7 = 5$ f) $16 + 3 \, x \, 2 + 5 = 37$

g) $24 / 4 + 2 \times 7 = 28$ h) $240 : 5 + 7 - 4 \, x \, 3 = 8$

66. Calculate:

a) $30 - 4 \cdot (5 + 2)$

b) $5 + 3 \cdot (8 - 6)$

c) $5 \cdot (11 - 3) + 7$

d) $3 \cdot (2 + 5) - 13$

e) $2 \cdot (7 + 5) - 3 \cdot (9 - 4)$

f) $4 \cdot (7 - 5) + 3 \cdot (9 - 7)$

g) $3 \cdot 5 - 3 \cdot (10 - 4 \cdot 2)$

h) $2 \cdot 3 + 5 \cdot (13 - 4 \cdot 3)$

Review exercises

1. Write the word names of the following numbers:

a) 50,038   b) 27,904,000   c) 2,187,000,000,600

2. Write these numbers as digits:

a) One thousand seven

b) Four hundred one thousand and ninety-six

c) Thirty-eight billion six hundred thousand

d) One hundred two million seventy thousand

e) Three billion twenty-three thousand five million

f) Sixty billion twenty-eight thousand and forty.

3. a) How many units are 7 tens of thousands?   b) How many tens are 6 units of million?

c) How many units in 24 hundreds?

d) How many hundreds are in 3 dozen thousand million?

4. Write the word names of the <u>ordinals</u> of:

a) 46   b) 138   c) 164   d) 22   e) 112   f) 81   g) 195   h) 211

5. Translate these romans into decimal system:

a) XIV   b) LXXIII   c) LXIX   d) CCXVII   e) DCXC   f) MCMLVI

6. Translate these romans into decimal system:

a) XXXIV   b) XCVI   c) MMCDXLIII   d) $\overline{VII}$ CCCIX

e) $\overline{XL}$ LXXVIII   f) $\overline{IX}$ g) $\overline{XXX}$ DCCXC

7. Write in romans:

    a) 18   b) 36   c) 54   d) 333   e) 608   f) 2 390

8. Write in romans:

    a) 29   b) 44   c) 699   d) 2,989   e) 3,479   f) 6,049

9. Calculate:

    a) $6\ 070 + 893 + 527$ b) $651 + 283 - 459$ c) $831 - 392 - 76$ d) $1\ 648 - 725 - 263$

10. Complete:

    a) $48 + \boxed{} = 163$ b) $\boxed{} + 256 = 359$ c) $628 - \boxed{} = 199$ d) $\boxed{} - 284 = 196$

11. Calculate without paper and pencil, as quickly as you can:

    a) $5 + 7 - 3 - 4$ b) $18 - 4 - 5 - 6$ c) $10 - 6 + 3 - 7$

    d) $8 + 5 - 4 - 3 - 5$ e) $12 + 13 + 8 - 23$ f) $40 - 18 - 12 - 6$

12. Calculate:

    a) $15 - 6 + 8$ b) $15 - (6 + 8)$ c) $12 - 7 - 2$ d) $12 - (7 - 2)$

    e) $27 - 11 + 12$ f) $27 - (11 + 12)$ g) $54 - 22 - 16$ h) $54 - (22 - 16)$

13. Calculate and check solutions:

    a) $18 - (6 + 9 - 3)$ b) $25 - (18 - 7) + 4$ c) $24 - (6 + 5 + 11)$

    d) $19 - (11 - 7) - 5$ e) $(26 - 17) + (32 - 24)$ f) $(33 - 25) - (24 - 19)$

    g) $(12 + 11) - (15 + 7)$ h) $(22 - 9) - (19 - 13)$

14. Calculate and check solutions:

    a) $5 - [7 - (2 + 3)]$ b) $3 + [8 - (4 + 3)]$ c) $2 + [6 + (13 - 7)]$

    d) $7 - [12 - (2 + 5)]$ e) $20 - [15 - (11 - 9)]$ f) $15 - [17 - (8 + 4)]$

15. Calculate:

    a) $16 \cdot 10$ b) $128 \cdot 10$ c) $60 \cdot 10$ d) $17 \cdot 100$ e) $85 \cdot 100$

    f) $120 \cdot 100$ g) $22 \cdot 1\ 000$ h) $134 \cdot 1\ 000$ i) $140 \cdot 1\ 000$

16. Calculate quotients and remainders:

a) 2647 : 8   b) 1345 : 29   c) 9045 : 45   d) 7482 : 174

e) 7971 : 2657   f) 27178 : 254

17. Copy and complete:

a) $123 \cdot \boxed{} = 5904$ b) $\boxed{} \cdot 86 = 1548$ c) $\boxed{} : 57 = 26$ d) $1\,862 : \boxed{} = 133$

18. Calculate a, b, c and d:

Dividend	Divisor	Quotient	Remainder
856	38	a	b
c	42	57	33
7512	d	156	24

19. Calculate:

a) $38 + 9 + 6419 + 80$ c) $89070 \cdot 80060$ e) $74895 - 8179$ g) $38259 : 46$

b) $6071 - 928$ d) $58264 : 7$ f) $69530 \cdot 90086$ h) $728603 : 379$

20. Calculate:

a) $8 + 7 - 3 \cdot 4$ b) $15 - 2 \cdot 3 - 5$ c) $22 - 6 \cdot 3 + 5$ d) $36 - 8 \cdot 4 - 1$

e) $4 \cdot 7 - 13 - 2 \cdot 6$ f) $5 \cdot 4 + 12 - 6 \cdot 4$ g) $5 \cdot 6 - 4 \cdot 7 + 2 \cdot 5$ h) $8 \cdot 8 - 4 \cdot 6 - 5 \cdot 8$

21. Calculate:

a) $2 \cdot (4 + 6)$ b) $2 \cdot 4 + 6$ c) $8 : (7 - 5)$ d) $5 \cdot 7 - 5$

e) $(5 + 6) \cdot 4$ f) $5 + 6 : 3$ g) $(19 - 7) : 2$ h) $18 - 7 \cdot 2$

22. Calculate and check your solution:

a) $30 - 4 \cdot (5 + 2)$ b) $5 + 3 \cdot (8 - 6)$ c) $5 \cdot (11 - 3) + 7$ d) $3 \cdot (2 + 5) - 13$

e) $2 \cdot (7 + 5) - 3 \cdot (9 - 4)$ f) $4 \cdot (7 - 5) + 3 \cdot (9 - 7)$

g) $3 \cdot 5 - 3 \cdot (10 - 4 \cdot 2)$ h) $2 \cdot 3 + 5 \cdot (13 - 4 \cdot 3)$

23. Calculate the value of the following expressions:

a) $6 + 4 \cdot 9$ b) $23 - 2 \cdot (5 + 3)$ c) $48 - 8 \cdot 5 - 10 + 18 : 3$

d) $7 + 5 \cdot (9 - 4) - 42 : 14$ e) $8 : 2 - 9 + (9 - 6 + 5) \cdot 4$ f) $15 - (8 - 5) : 3 - 3 \cdot (6 + 5) + 2 \cdot 4 \cdot 7$

g) $30 - 2 \cdot [26 - (15 - 3)]$ h) $100 + 10 \cdot (45 \cdot 2 - 80) \cdot 10$ i) $6 - 2 \cdot (5 + 8 : 4) + (9 - 2 \cdot 3) : 3$

j) $8 + 7 \cdot (6 + 4 \cdot 5) - [5 \cdot (25 + 5) - 10 \cdot (3 - 2)]$

24. In a whole division, divisor is 54, quotient is 25 and remainder is 12. Calculate the dividend.

25. Ana has difficulties to distribute 50 kg of apples in boxes of 3 kg. How can she do it?

26. a) How many 30 € balls you can buy with 250 €? How much money will you have left?

b) How many 60 € balls you can buy with 500 €? How much money will you have left?

c) How many 15 € balls you can buy with 125 €? How much money will you have left?

d) What do you observe when comparing the results of the three previous sections? Why?

27. An employee earned, in January, 2,056 €, in February, 136 € less, and in March, 287 € plus than in February. How much did he earn during the first quarter?

28. Adela had 1187 € in the bank, but he had paid 385 € with her credit card for a coat and 163 for a dress. How much does she have left?

29. The median goose weighs 850 g more than the small and 1155 g less than great. What is their total weight?

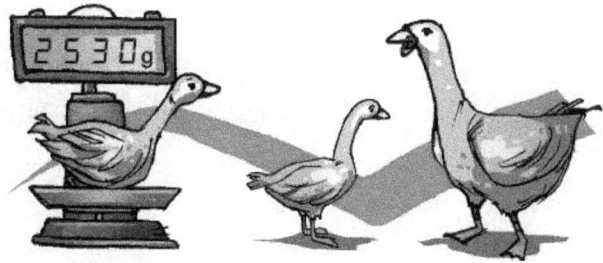

30. In an international marathon, there are 187 European runners, 145 Americans and 158 Asians. The rest, up to a total of 612 participants are African. How many of the runners are African?

31. The fence wall of my school has eight bars per meter, and has a length of 327 meters. How many bars are there?

32. You want to plant trees, spaced 20 meters along a path that has a length of two kilometers. How many trees are needed?

33. A farmer has 483 sheeps. If the average value of each sheep is 87 €, what is the total value of the whole flock?

34. A truck has traveled 450 km in 6 hours. What is its mean speed?

35. A walker walks at a rate of 72 steps per minute, advancing 85 cm per step. How far will he be in an hour?

36. A car factory has produced 15,660 units in the last three months. How many cars are produced, on average, every day?

37. A fishing boat has got 9100 € for a capture of 1300 kg. How much will another boat coming into port with 1750 kg of the same product and quality get?

38. A farmer has sold today 85 kg of tomatoes and 35 kg of strawberries in a market. If he sells tomatoes at 2 €/kg and strawberries at 3 €/kg, how much will he get?

39. A delivery truck is carrying 15 boxes of orange soda and 12 boxes of lemon soda. How many bottles does it take if each box contains 24 units?

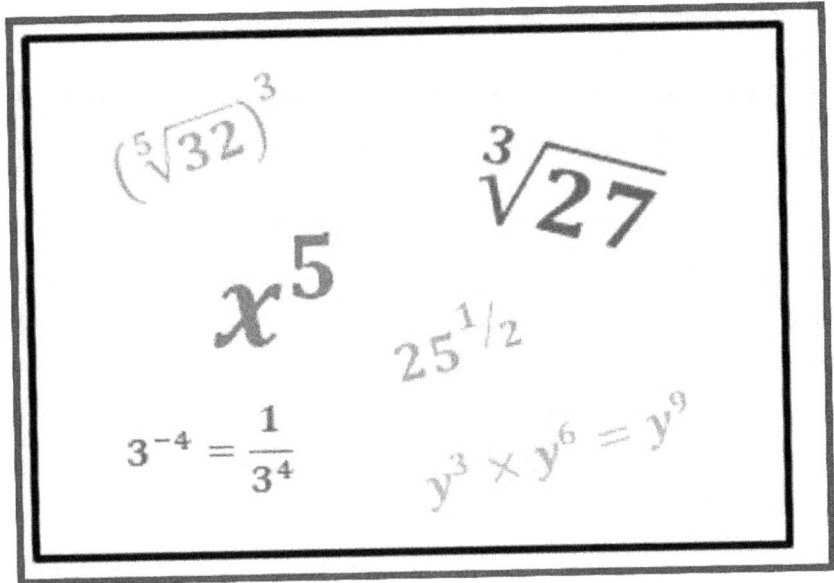

Unit 2.- Powers and roots

1. Powers

A **Power** is a number obtained by multiplying a number by itself a certain number of times

$$a^n = \underbrace{a \cdot a \cdot a \cdot a \cdots a}_{n \; times} \quad where \quad \begin{cases} a : base \\ b : \exp onent \end{cases}$$

For example, $6^5 = 6 \cdot 6 \cdot 6 \cdot 6 \cdot 6$, and can be read as:

- Six to the fifth power

- Six to the power of five

- Six powered to five.   The most common is **six to the power of five.**

The number that is successively multiplied by itself is called the **base**. A small raised number called the **exponent** follows the base and indicates the number of times the base is to be multiplied.

And, remember: $\boxed{a^1 = a}$ and $\boxed{a^0 = 1}$, for any natural number a.

Especial cases: Squares and cubes (powers of two and three)

Although 3^2 is read as three to the second power, three to the power of two, three to the square power, the most common is **three squared.**

Although 5^3 is read as five to the third power, five to the power of three, the most common is **five cubed.**

1. Express as a power the following products:

$5 \cdot 5 \cdot 5 \cdot 5 =$ $\qquad$ $3 \cdot 3 =$

$7 =$ $\qquad$ $(-4)(-4)(-4)(-4)(-4) =$

$(-1)(-1)(-1) =$ $\qquad$ $0 \cdot 0 \cdot 0 \cdot 0 \cdot 0 \cdot 0 \cdot 0 =$

$2 \cdot 2 \cdot 2 \cdot 2 \cdot 2 \cdot 2 =$ $\qquad$ $1 \cdot 1 \cdot 1 \cdot 1 \cdot 1 =$

$(-5)(-5) =$ $\qquad$ $(-9)(-9)(-9)(-9) =$

$8 \cdot 8 \cdot 8 =$ $\qquad$ $(-3)(-3)(-3)(-3) =$

2. Express the following powers as products:

a) $3^4 =$ $\quad$ b) $2^7 =$ $\quad$ c) $9^3 =$ $\quad$ d) $15^2 =$ $\quad$ e) $10^6 =$ $\quad$ f) $20^4 =$

3. Copy and complete the gab indicated with symbol "Δ":

a) $m \cdot m \cdot m \cdot m = m^\Delta$ $\qquad$ b) $x \cdot x = x^\Delta$ $\qquad$ c) $5 \cdot 5 \cdot 5 \cdot 5 = \Delta^4$

4. Calculate:

a) 5^3 $\qquad$ b) 2^6 $\qquad$ c) 4^4 $\qquad$ d) 10^3

5. Calculate the value of "x":

a) $3^x = 27$ b) $5^x = 25$ c) $2^x = 16$ d) $7^x = 343$

6. Calculate the value of "x":

a) $x^4 = 81$ b) $x^3 = 125$ c) $x^2 = 36$ d) $x^7 = 128$

7. Calculate:

a) 2^2 b) 9^2 c) 15^2 d) 2^3 e) 7^3 f) 11^3

8. Continue the sequence of the squares of the natural numbers: 0, 1, 4, 9…

9. Calculate:

a) The squared of 100   b) The cube of 10   c) The squared of 20   d) The cube of 6.

10. Express with all its digits:

a) 10^7 b) 10^{10} c) 10^{15} d) 10^1

11. Calculate the value of "x":

a) $10^x = 100$ b) $10^x = 10\,000$ c) $10^x = 1\,000\,000$ d) $10^x = 100\,000\,000$

12. Write the numbers whose polynomial decomposition is:

a) $3 \cdot 10^4 + 8 \cdot 10^3 + 5 \cdot 10^2 + 7 \cdot 10 + 9$ b) $2 \cdot 10^6 + 7 \cdot 10^4 + 6 \cdot 10^3 + 5 \cdot 10 + 4$

c) $8 \cdot 10^6 + 5 \cdot 10^5$ d) $4 \cdot 10^9 + 3 \cdot 10^8 + 6 \cdot 10^7$

13. Write the polynomial decomposition of:

a) 7 294   b) 5 238 427   c) 37 250 000   d) 30 800 050

14. Express in an abbreviated way these quantities:

a) Number of cells in a human mean body is 25 000 000 000

b) Number of molecules in a liter of water is 334 326 000 000 000 000 000

15. Write with all its digits:

a) $8 \cdot 10^5$ b) $54 \cdot 10^4$ c) $16 \cdot 10^9$ d) $37 \cdot 10^{13}$

29

2. Operations with powers

2.1. Product of powers with the same base

When powers with the same base are multiplied, the base remains unchanged and the **exponents are added.**

$$a^m \cdot a^n = a^{(m+n)}$$

16. Write as an only power:

a) $2^4 \cdot 2^5$ b) $3^2 \cdot 3^3$ c) $2^4 \cdot 2^2$ d) $5^2 \cdot 5^6$ e) $7 \cdot 7^3$ f) $4^0 \cdot 4^2$ g) $2^6 \cdot 2$ h) $5^3 \cdot 5^2$

17. Find the value of x that makes the following expressions true:

a) $2^x \cdot 2^4 = 2^6$

b) $3^2 \cdot 3^x \cdot 3^4 = 3^{10}$

c) $2^8 \cdot 2^x \cdot 2^{15} = 2^{28}$

d) $3^4 \cdot 3^7 \cdot 3^8 \cdot 3^{15} = 3^x$

e) $2^4 \cdot 2^5 \cdot 2 = 2^x$

f) $2^3 \cdot 3^2 \cdot 3^4 \cdot 2^5 \cdot 3^6 = 2^x \cdot 3^y$

g) $2^x \cdot 2^x \cdot 2^5 = 2^7$

h) $7^x \cdot 7^2 \cdot 7^x = 7^{12}$

2.2. Quotient of powers with the same base

When powers with the same base are divided, the base remains unchanged and the **exponents are subtracted.**

$$\frac{a^m}{a^n} = a^{(m-n)}$$

18. Write as an only power:

a) $2^5 : 2^2$

b) $3^7 : 3^3$

c) $2^5 : 2$

d) $5^5 : 5^4$

19. Find the value of x that makes the following expressions true:

a) $2^x : 2^4 = 2^6$

b) $3^6 : 3^x = 3^2$

c) $2^8 : 2^x = 2^5$

d) $3^4 : 3^x = 3$

Exercises

30

2.3. Power of a power

The exponents must be **multiplied**

$$\left(a^m\right)^n = a^{m \cdot n}$$

Exercises

20. Calculate:

a) $(2^3)^2$ b) $(3^2)^4$ c) $((5^3)^2)^4$

21. Fill in the missing numbers:

a) $(4^2)^5 = 4^{[\ ]}$ b) $(3^2)^{[\ ]} = 3^8$ c) $(3^{[\ ]})^2 = 3^{10}$ d) $([\ ]^2)^3 = 5^6$ e) $(2^2)^{[\ ]} = [\ ]^8$

2.4. Powers with different bases but the same exponent

Product: When powers **with the same exponent** are multiplied, multiply the bases and keep the **same exponent**.

$$(a \cdot b)^m = a^m \cdot b^m$$

Quotient: When powers **with the same exponent** are divided, bases are divided and the **exponent remains unchanged**.

$$\left(\frac{a}{b}\right)^m = \frac{a^m}{b^m}$$

Exercises

22. Complete:

a) $3^7 \cdot 8^7 = [\ ]^7$ b) $[\ ]^2 \cdot [\ ]^2 = 6^2$ c) $5^2 \cdot [\ ]^2 = 15^2$ d) $2^2 \cdot [\ ]^2 = 14^2$

e) $8^5 : 4^5 = [\ ]^5$ f) $\dfrac{15^5}{5^5} = [\ ]^5$ g) $\dfrac{16^7}{8^7} = [\ ]^7$ h) $\dfrac{6^{12}}{3^{12}} = [\ ]^{[\ ]}$

Operations with powers

Laws	Examples
$a^m \cdot a^n = a^{(m+n)}$	$a^4 \cdot a^3 = (a \cdot a \cdot a \cdot a) \cdot (a \cdot a \cdot a) = a \cdot a \cdot a \cdot a \cdot a \cdot a \cdot a = a^7$
$\dfrac{a^m}{a^n} = a^{(m-n)}$	$\dfrac{a^8}{a^3} = \dfrac{a \cdot a \cdot a \cdot a \cdot a \cdot \not{a} \cdot \not{a} \cdot \not{a}}{\not{a} \cdot \not{a} \cdot \not{a}} = a \cdot a \cdot a \cdot a \cdot a = a^5$
$\left(a^m\right)^n = a^{m \cdot n}$	$(a^4)^2 = a^4 \cdot a^4 = a \cdot a \cdot a \cdot a \cdot a \cdot a \cdot a \cdot a = a^8$
$(a \cdot b)^m = a^m \cdot b^m$	$(a \cdot b)^3 = (a \cdot b) \cdot (a \cdot b) \cdot (a \cdot b) = a \cdot b \cdot a \cdot b \cdot a \cdot b = (a \cdot a \cdot a) \cdot (b \cdot b \cdot b) = a^3 \cdot b^3$
$\left(\dfrac{a}{b}\right)^m = \dfrac{a^m}{b^m}$	$\left(\dfrac{a}{b}\right)^3 = \dfrac{a}{b} \cdot \dfrac{a}{b} \cdot \dfrac{a}{b} = \dfrac{a \cdot a \cdot a}{b \cdot b \cdot b} = \dfrac{a^3}{b^3}$

Exercises

22. Write as an only power:

a) $3^2 \cdot 3^2$ b) $2^3 \cdot 2^5$ c) $4^3 \cdot 4^5$ d) $10^5 \cdot 10^2$ e) $3 \cdot 3^2 \cdot 3^3$ f) $5 \cdot 5^4 \cdot 5^4$

23. Write as an only power:

a) $2^6 : 2^2$ b) $3^8 : 3^5$ c) $4^7 : 4^6$ d) $10^5 : 10^3$ e) $(7^5 : 7^3) : 7^2$ f) $(5^9 : 5^4) : 5^3$

24. Write as an only power:

a) $(5^2)^3$ b) $(2^6)^2$ c) $(3^2)^2$ d) $(4^3)^4$ e) $(7^2)^4$ f) $(5^4)^2$

25. Reduce:

a) $a^3 \cdot a^5$ b) $a^8 : a^6$ c) $(a^3 \cdot a^6) : a^5$ d) $(a^{10} : a^7) : a^2$ e) $(a^2)^5 : (a^3)^2$ f) $(a^4)^3 : (a^6)^2$

26. Write as an only power:

a) $8^5 : 4^5$ b) $12^3 : 4^3$ c) $5^3 \cdot 2^3$ d) $25^2 \cdot 4^2$ e) $(6^4 \cdot 3^4) : 9^4$ f) $(2^5 \cdot 3^5) : 6^5$

27. Write as an only power:

a) $2^3 \cdot 2^4$ b) $3^5 \cdot 3^3 \cdot 3^4$ c) $7^2 \cdot 7^5$ d) $4^5 : 4^2$ e) $7^6 : 7^3$

f) $x^9 \cdot x^4$ g) $(5^2)^3$ h) $(2^2)^3$ i) $2^9 : 2^5$ j) $9^7 : 9^4$

28. Write as an only power:

   a) $5^4 \cdot 3^4$ b) $2^2 \cdot 4^2$ c) $3^4 \cdot 4^4 \cdot 5^4$ d) $x^n \cdot y^n$ e) $5^4 \cdot 3^4$

   f) $8^3 : 4^3$ g) $9^5 : 3^5$ h) $x^n : y^n$ i) $10^4 : 2^4$ j) $6^3 : 3^3$

29. Calculate by using the laws of powers:

   a) $2^3 \cdot 5^4$ b) $(6^5 : 2^4) : 3^5$ c) $\dfrac{20^6}{2^6}$ d) $\dfrac{20^6}{2^5}$

---

3. Roots

Till now, we have calculated and worked with powers. In this part of the unit, we a are going to work with the function "root", which is reciprocal to a power.

For example, if we say that $\sqrt{9} = 3$, that means that $3^2 = 9$.

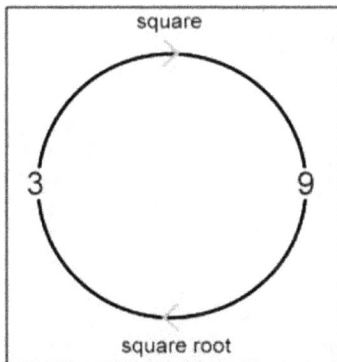

Examples:

$\sqrt{1}=1$ because $1^2=1$
$\sqrt{4}=2$ because $2^2=4$
$\sqrt{9}=3$ because $3^2=9$
$\sqrt{16}=4$ because $4^2=16$
$\sqrt{25}=5$ because $5^2=25$
$\sqrt{36}=6$ because $6^2=36$
$\sqrt{49}=7$ because $7^2=49$

Or, in another way:

POWERS	ROOTS
A square has a side of 5 cm. What is its area? $A = s^2 = 5^2 = 25$ cm^2.	The area of a square is 49 cm^2. What is the length of its side? $s = 7$ cm, because $7^2 = 49$. This is a root, a square root: $\sqrt{49} = 7 \iff 7^2 = 49$

> **Squared root** of a real number **a,** written $\sqrt{a}$, is another real number, **b,** so that $b^2 = a$.
>
> $$\sqrt{a} = b \iff b^2 = a$$
>
> **a** is named **radicand**, and **b** is named **square root**.

Number of roots

- If radicand is a **positive** number, there are **two opposite roots**.

    For example, squared root of 25 is 5 and (-5).

- If radicand is **0**, its root is **0.**

- If radicand is a **negative** number, it has **no roots,** because any number to an even power, is always a positive number.

    For example, (-4) has no roots, because there is not any number b, so that b2 = (-4).

Radicand: a	Number of roots
a > 0	2 opposite roots
a = 0	1 null root
a < 0	No roots

Exact and whole roots

Sometimes, when calculating square root of a number, we exactly obtain the solution, like $\sqrt{25} = 5$. This is named an **exact root**.

But, it is more usual not to obtain an exact solution, but having a remainder. For example, when calculating square root of 40, we try squares: $3^2 = 9$, $4^2 = 16$, $5^2 = 25$, $6^2 = 36$, $7^2 = 49$.

As $6^2 < 40 < 7^2$, we say 40 has a whole square root of 6, having a remainder of $40 - 36 = 4$.

So, we say 40 has a **whole square root** of 6 and a **remainder** of 4.

Example 1: Calculate whole root and remainder of 11, 29 and 37.

<u>Solution:</u>

a) $2^2 = 4$, $3^2 = 9$, $4^2 = 16 > 11$ So, whole root is 3 and remainder is $11 - 3^2 = 11 - 9 = 2$.

b) Whole root = 5,   Remainder = $29 - 5^2 = 29 - 25 = 4$.

c) Whole root = 6,   Remainder = $37 - 6^2 = 37 - 36 = 1$.

30. Verify if the following square roots are true:

a) $\sqrt{121} = 11$ b) $\sqrt{196} = 14$ c) $\sqrt{300} = 30$ d) $\sqrt{10000} = 100$

31. Calculate the length of the side of a square whose area is 81 m^2.

A=81

32. Find the whole square root and the remainder of the numbers:

a) 15   b) 39   c) 57   d) 95

33. Complete:

a) $\sqrt{50} \approx 7$ Remainder =   b) $\sqrt{\underline{}} \approx 5$ Remainder = 7

c) $\sqrt{8} \approx 2$ Remainder =   d) $\sqrt{\underline{}} \approx 9$ Remainder = 2

Exercises

Review exercises

1. Write as a power:

   a) $x^8 : x^7$ b) $y^5 \cdot y^7$ c) $(z^2)^4$ d) $(x^3)^3$ e) $y^5 : y^3$

   f) $z^9 \cdot z$ g) $x^8 \cdot x^0$ h) $(y^0)^3$ i) $z^9 : z^9$

   Sol.: a) x b) y^{12} c) z^8 d) x^9 e) y^2 f) z^{10} g) x^8 h) $y^0 = 1$ i) $z^0 = 1$

2. Simplify to an only power:

   a) $7^4 \cdot 7^5$ b) $5^3 \cdot 5^3$ c) $9^3 \cdot 9^5 \cdot 9^4$ d) $4^2 \cdot 4^3 \cdot 4^4$ e) $7^8 : 7^5$

   f) $20^6 : 20^6$ g) $9^7 : 9^4 : 9^2$ h) $6^{10} : 6^5 : 6^2$ i) $(5^4)^2$ j) $(7^3)^2$

   k) $(5^0)^2$ l) $(4^1)^3$ m) $(3^5 \cdot 3^2) : 3^3$ n) $4^3 \cdot (4^7 : 4^4)$ ñ) $(8^5 : 8^3) \cdot 8^2$

   o) $7^5 : (7^2 \cdot 7^2)$ p) $(3^5)^2 \cdot (3^2)^4$ q) $(7^3)^3 \cdot (7^2)^4$ r) $(9^5)^3 : (9^4)^3$ s) $(2^9)^2 : (2^3)^5$

     Sol.:   a) 7^9 b) 5^6 c) 9^{12} d) 4^9 e) 7^3 f) 1 g) 9 h) 6^3 i) 5^8 j) 7 k) 1 l) 4^3

                            m) 3^4 n) 4^6 ñ) 8^4 o) 7 p) 3^1 q) 7^{17} r) 9^3 s) 2^3

3. Calculate:

   a) $(5^3 \cdot 4^3) : 2^3$ b) $6^3 : (21^3 : 7^3)$ c) $36^4 : (2^4 \cdot 9^4)$ d) $(2^4 \cdot 2^5) : 2^9$

   e) $(15^5 : 5^5) : 3^3$ f) $12^9 : (4^7 \cdot 3^7)$ g) $(4^3 \cdot 4^5) : (4^4 \cdot 4^2)$ h) $(30^7 : 5^7) : (2^5 \cdot 3^5)$

   Sol.: a) $10^3 = 1\,000$ b) $2^3 = 8$ c) $2^4 = 16$ d) $2^0 = 1$ e) $3^2 = 9$ f) $12^2 = 144$ g) $4^2 = 16$

                                                       h) $6^2 = 36$

4. Simplify to only one power:

   a) $(a^3 \cdot a^2) : a^4$ b) $(x^5 : x) \cdot x^2$ c) $(m^7 : m^4) : m^3$ d) $(a^3)^4 : a^{10}$

   e) $(x^2)^4 : (x^2)^3$ f) $(m^4)^3 : (m^5)^2$ g) $(a^3 \cdot a^5) : (a \cdot a^4)$ h) $(x^3 : x^2) \cdot (x^4 : x^3)$

                     Sol.: a) a^1 b) x^5 c) $m^0 = 1$ d) a^2 e) x^2 f) m^2 g) a^3 h) x^2

5. Simplify to only one power and calculate:

Clue: $9 = 3^2$ *That means, for example:* $9^5 = (3^2)^5 = 3^{10}$

---

   a) $2^{10} : 4^4$ b) $3^6 : 9^2$ c) $25^3 : 5^4$ d) $(2^3 \cdot 4^2) : 8$ e) $(3^4 \cdot 9^2) : 27^2$

   f) $(5^5 \cdot 5^3) : 25^3$ **Sol.:** a) 4 b) 9 c) 25 d) 16 e) 9 f) 25

6. Copy and complete as in the example: $8^2 = 64 \leftrightarrow \sqrt{64} = 8$

a) $\boxed{\cdots}^2 = 36 \leftrightarrow \sqrt{36} = \boxed{\cdots}$ b) $\boxed{\ldots} = 256 \leftrightarrow \sqrt{256} = \boxed{\ldots}$ **Sol.:** a) 6 b) 16

7. Find out the value of *m*:

a) $\sqrt{m} = 8$ b) $\sqrt{m} = 20$ c) $\sqrt{m} = 45$ **Sol.:** a) 64 b) 400 c) 2025

8. Find out the value of *a*:

a) $a^2 = 81$ b) $a^2 = 100$ c) $a^2 = 441$ **Sol.:** a) 9; b) 10; c) 21.

9. How many tiles are needed to cover a square room of side 22 m, if each tile has a surface of 1 square meter? *Clue: tile = losa / azulejo.*   **Sol.:** 484 tiles.

10. A square farm has an area of 900 square meters. Calculate the length of its side. **Sol.:** 30 meters.

11. How many tiles do you need to cover a square wall, if the first row has 5 tiles?   **Sol.:** 25 tiles.

12. The water deposit of town is a cube and its edge is 12 meters. Calculate the reservoir volume expressing the result as a power. *Clue: edge = arista.*   **Sol.:** 1728 m^3.

13. An NGO sends humanitarian 8 trucks with 8 identical boxes each. In each box there are 8 packets. Every packet contains 8 bags with 8 blankets each. How many blankets are sent in total? Write it as a power. *Clue: NGO = ONG, blanket = manta.*   **Sol.:** 8^5 blankets.

14. Complete:

a) $\sqrt{16} = \boxed{} \Rightarrow 4^2 = \boxed{}$ d) $\sqrt{32} = 2^{\boxed{}} \Rightarrow \boxed{}^5 = \boxed{}$

b) $\sqrt[3]{\boxed{}} = 2 \Rightarrow \boxed{}^3 = \boxed{}$ e) $\sqrt[3]{\boxed{}} = 3 \Rightarrow 3^3 = \boxed{}$

c) $\sqrt[4]{16} = \boxed{} \Rightarrow 2^{\boxed{}} = \boxed{}$ f) $\sqrt{\boxed{}} = 9 \Rightarrow \boxed{}^2 = 81$

15. Answer the following questions:

    a) Calculate the area of a 9 m side.

    b) If the side of a square is 14 cm, what is its area?

    c) Calculate the side of a square if its area is 64 m^2.

    d) If the area of a square is 169 cm^2, how long is its side?

16. How many square shaped tiles of side 20 cm are needed to cover a square shaped room of 1 m side?

If you want to review this unit

17. Calculate:

    a) 2^7 b) 3^5 c) 5^3 d) 10^3 e) 1^{16} f) 1^{17}

                                 Sol.: a) 128; b) 243; c) 125; d) 1000; e) 1; f) 1.

18. Express as an only power:

    a) $2^4 \cdot 2^3$ b) $2^3 \cdot 2^3$ c) $3^5 : 3^3$ d) $5^6 : 5^3$ **Sol.:** a) 2^7; b) 2^6; c) 3^2.

19. Calculate:

    a) $2^3 + 3^3 - 4^2$ b) $5^2 \cdot 2^2 + 3^2 \cdot 3$ c) $2^2 \cdot [5^2 - 4^2]$ d) $6^3 : 3^3 + 8^2 : 4^2$

                                       Sol.: a) 19; b) 127; c) 36; d) 12.

20. Express as a power:

    a) $6 \cdot 6 \cdot 6 \cdot 6$ b) $5 \cdot 5 \cdot 5$ c) $2 \cdot 2 \cdot 2 \cdot 2 \cdot 2$ **Sol.:** a) 6^4; b) 5^3; c) 2^5.

21. Express as an only power:

    a) $3^3 \cdot 3 \cdot 3^5$ b) $5^2 \cdot 5^5 \cdot 5$ **Sol.:** a) 3^9; b) 5^8.

22. Express as an only power:

    a) $3^8 : 3^2$ b) $5^6 : 5$ **Sol.:** a) 3^6; b) 5^5.

23. Express as an only power:

    a) $(3^2)^5$ b) $(5^6)^2$ **Sol.:** a) 3^{10}; b) 5^{12}.

24. Express as an only power:

a) $(2 \cdot 5)^4$ b) $(5 \cdot 4 \cdot 2)^6$ **Sol.:** a) 10^4; b) 40^6.

25. Write as a power of ten:

a) 1000   b) 10 000   c) 1 000 000   d) 100 000 000

Sol.: a) 10^3; b) 10^4; c) 10^6; d) 10^9.

26. Express as an only power:

a) $3^4 \cdot 4^3 \cdot 4$ b) $5^4 \cdot 5^{30} \cdot 5^2$ c) $4^3 \cdot 4^2 \cdot 4$ d) $2^4 \cdot 2^5$

Sol.: a) 12^4; b) 5^{36}; c) 4^4; d) 2^9.

27. Express as an only power:

a) $2^4 \cdot 2^3$ b) $3^4 \cdot 3^6$ c) $5^6 : 5^2$ **Sol.:** a) 2^7; b) 3^{10}; c) 5^4.

28. Calculate:

a) 10^3 b) 10^4 c) 10^5 d) 10^6 **Sol.:** a) 1000; b) 10 000; c) 100 000; d) 1000 000.

29. Write with all its digits:

a) $24 \cdot 10^7$ b) $5 \cdot 10^8$ c) $43 \cdot 10^5$ **Sol.:** a) 240 000 000; b) 500 000 000; c) 4 300 000.

30. Write in an abbreviated way. Look at the example:

a) $27\ 000\ 000 = 27 \cdot 10^6$ b) 30 000 000 000   c) 2 300 000

Sol.: b) $3 \cdot 10^{10}$; c) $23 \cdot 10^5$.

31. Calculate:

a) $\sqrt{196}$ b) $\sqrt{441}$ c) $\sqrt{225}$ **Sol.:** a) 14; b) 21; c) 15.

32. Calculate the square root of the following numbers

a) 100   b) 49   c) 81   d) 121   **Sol.:** a) 10; b) 7; c) 9; d) 11.

33. Write the two naturals between which following squared rotor are:

a) ____ $< \sqrt{20} <$ ____   b) ____ $< \sqrt{40} <$ ____

c) ____ $< \sqrt{90} <$ ____   d) ____ $< \sqrt{140} <$ ____

Sol.: a) 4 and 5; b) 6 and 7; c) 9 and 10; d) 11 and 12.

34. To cover a square shaped room, we needed, exactly, 169 tiles. Find out:

a) How many tiles did we put in each row?

b) If each tile has a side of 40 cm, How many square shaped meters does the room have?

Sol.: a) 13; b) 270,400.

35. Calculate the whole root and the remainder of the following numbers:

a) 97   b) 134   c) 1500

Sol.: a) 9 and 16; b) 11 and 13; c) 38 and 56.

36. Calculate the whole root and the remainder of the following numbers:

a) 2560   b) 13456

Sol.: a) 50 and 60; b) 116 and 0 (exact).

Unit 3.- Divisibility

1. Multiples and divisors

A *multiple* of a whole number is the product of that number and a natural number. Also, a natural number is a multiple of another if it is in its multiplication tables.

So, we can see it in two ways:

For example:
- multiples of 2 are 2, 4, 6, 8, …
- multiples of 3 are 3, 6, 9, 12, …
- multiples of 4 are 4, 8, 12, 16, …

For example:
- 21 is a multiple of 7 because 7 x 3 = 21
- 77 is a multiple of 7 because 7 x 11 = 77
- 98 is a multiple of 7 because 7 x 14 = 98
- 147 is a multiple of 7 because 7 x 21 = 147.

A whole number that divides exactly another whole number is called a ***divisor*** of that number.

For example 20 : 4 = 5.   So, 4 is a divisor of 20.

Notice: If a number can be expressed as a product of two whole numbers, then these whole numbers are *divisors* of that number.

Exercises

1. Indicate if there is a divisibility relationship between these pairs of numbers:

    a) 294 and 14   b) 360 and 15   c) 115 and 15   d) 561 and 17

2. Explain in a clear way why 184 is a multiple of 23.

3. Is 17 a divisor of 255? Why?

4. Find out three numbers being multiples of 25.

5. Find out three divisors of 30.

6. a) Is 200 a divisor of 1 000?   Is 1 000 a multiple of 200?

    b) Is 15 a divisor of 70? Is 70 a multiple of 20?

    c) Is 90 a multiple of 6?

    d) Is 125 a divisor of 1000?

7. Write: a) Five multiples of 6.   b) Five multiples of 17.   c) Five multiples of 200.

8. Which of these numbers are multiples of 9?

81   16   53   36   99   108   44

9. Add four terms to each sequence:

    a) 3, 6, 9, 12, …   b) 13, 26, 39, 52, …   c) 15, 30, 45, 60, …   d) 51, 102, 153, 204, …

10. Write the ten first multiples of 15.

11. Find out all the multiples of 8 being between 700 and 750.

12. Write the first multiple of 31 being bigger than 1 000.

13. Write ten first multiples of 5. Look at the last digit. What do you observe? How could you know, only by looking at it, if a number is a multiple of 5?

14. Find a number that has all these divisors: 1, 2, 3, 4, 24, 12, 8, 6.

15. We want to pack 36 bottles into similar packs. In what ways can we do it?

16. Is there a divisibility relationship between these pairs of numbers?

a) 20 y 300   b) 13 y 195   c) 38 y 138   d) 15 y 75   e) 23 y 203

17. Answer the following questions:

a) Is 372 a multiple of 12? Why?   b) Is 21 a divisor of 189? Why?

18. Continue the following multiples´ serie: Multiples of 12:   12, 24, 36, __, __, __

19. Continue the following multiples´ serie: Multiples of 15:   15, 30, 45, __, __, __

20. Write down the five lowest multiples of 11.

21. Write down the multiples of 20 between 150 and 210.

22. Write down one multiple of 12 between 190 and 200.

---

2. Tests of divisibility

One number is divisible by:

2 If the last digit is 0 or an even (0, 2, 4, 6, 8).

3 If the sum of the digits is divisible by 3.

5 If the last digit is 0 or 5.

11 If the sum of the digits in the even position minus the sum of the digits in the odd position is 0 or divisible by 11.

23. Write the sequence of the 15 first multiples of 10. How could you know if a number is a multiple of 10 only by observing it?

24. Which of the following numbers are multiples of 3?

127   195   369   444   570   653   821   1302

25. Copy these numbers in your notebook:

138   150   153   285   299   356   375   400   412   515

a) Round multiples of 2 with a red pencil.   b) Round multiples of 5 with a blue pencil.

c) Which are the multiples of 10?

26. Find which of the numbers are multiples of:

239   300   675   570   800   495   888   6402   2088

a) 2:   b) 3:   c) 5:   d) 10:   e) 11:

27. Change every letter by the digit/s that make the number divisible by 3.

A51   2B8   31C   52D   1E8

28. For each number, change every letter by the digit/s that make the number divisible, at the same time, by 2 and 3.

4 a 3 2 a 2 4 a

29. Charlie left his coin collection to be divided evenly among his three children. His collection contains 439 coins. Is it possible for each child to receive the same number of coins? Explain.

30. Charlie's coin collection is valued at $126. Is it possible to divide the collection so that each of his three children receives the same dollar amount?

31. Pedro and two friends rent an apartment for $987 per month. Is it possible for each of them to spend the same whole number of dollars on the rent? Explain

32. Helen and four of her friends won the lottery. They received a check for $65,475. Is it possible for them to split the money evenly? Explain.

33. A marching band has 175 members. Can the band march in rows of 3 without any member being left over? Rows of 5? Rows of 10?

34. Karla and her five friends are driving 1362 miles to attend a wedding. Is it possible for each of them to drive the same number of miles?

….. And aren't there tests of divisibility for composite numbers?

Divisibility tests are only for studied prime numbers. For *composite* numbers, you can apply the following method:

Given a composite number A $(A = a \cdot b \cdot c \cdot .... \cdot j)$, another number B will be divisible by A if it is divisible, separately, by $a, b, c, ...$ and j.

Although 10 is a composite number ($10 = 2 \cdot 5$), we have studied a divisibility test for it. Check it verifies above statement.

Example 1: Check why we can know 4020 is divisible   a) by 6,   b) 15 and   c) 30.

Solution:

   a) Notice that $6 = 2 \cdot 3$,   4020 is divisible by 2 (last digit is an even number) and by 3 ($4 + 2 = 6$).

   b) Notice that $15 = 3 \cdot 5$, 4020 is divisible by 3 ($4 + 2 = 6$) and by 5 (last digit is 0).

   c) Notice that $30 = 2 \cdot 3 \cdot 5$,   4020 is divisible by 2, 3 and 5, so it is also divisible by 30.

35. Is 35,766 divisible by 6?

36. Is 11,370 divisible by 15? Write a divisibility test for 15.

37. Is 11,370 divisible by 30? Write a divisibility test for 30.

A *prime number,* or a *prime* is a whole number greater than 1 with exactly two factors (divisors). The two factors are the number 1 and itself.

A *composite number,* or a *composite,* is a whole number greater than 1 with more than two factors (divisors).

What about **0** and **1**?   0 and 1 are neither prime nor composite.

Exercises

38. Find out all the prime numbers less than 50.

39. Among the following numbers, there are two primes. Which are they?

29   49   59
39   69

The sieve of Eratosthenes

In one of the previous exercises, you have had to find out the primes less than 50. Could you think on a method to solve it in the shortest way?

A Greek mathematician, Eratosthenes (276 - 195 BC), discovered the sieve which is known as the Sieve of Eratosthenes. It is a method to get prime numbers. He got a table of whole numbers, e.g. from 1 to 100 and crossed out the number 1. After that, he crossed out number 2 and all the multiples of 2. Later, 3 and its multiples, and he went on in this way.

Example 2: Copy the following table in your notebook and build the Eratosthenes' sieve.

1	2	3	4	5	6	7	8	9	10
11	12	13	14	15	16	17	18	19	20
21	22	23	24	25	26	27	28	29	30
31	32	33	34	35	36	37	38	39	40
41	42	43	44	45	46	47	48	49	50
51	52	53	54	55	56	57	58	59	60
61	62	63	64	65	66	67	68	69	70
71	72	73	74	75	76	77	78	79	80
81	82	83	84	85	86	87	88	89	90
91	92	93	94	95	96	97	98	99	100

4. Prime factorization

The *prime factorization* of a natural number is the indicated product of prime numbers.

To find the prime factorization of a natural number using repeated division:

1. Divide the natural number and each following quotient by a prime number until the quotient is 1. Begin with 2 and divide until the quotient is odd, then divide by 3. Divide by 3 until the quotient is not a multiple of 3. Continue with 5, 7, and so on, testing the prime numbers in order.

2. Write the indicated product of all the divisors.

Example 3: Find the prime factorization of the number 180:

Solution:

```
180 | 2
 90 | 2
 45 | 3
 15 | 3
  5 | 5
  1 |
```

So, prime factorization of 180 is: $2^2 \cdot 3^2 \cdot 5$.

Exercises

40. Find the prime factorization of the following numbers:

| 6 | 8 | 10 | 14 | 15 | 18 |

| 20 | 24 | 25 | 27 | 30 | 42 |

41. Find the prime factorization of the following numbers:

a) 48   b) 54   c) 90   d) 105

e) 120   f) 135   g) 180   h) 200

42. Work out the prime factorization of the following numbers:

a) 12   b) 18   c) 24   d) 36   e) 42   f) 360   g) 1350   h) 48

43. What are the numbers that have the following prime factorization?

a) $2^3 \cdot 3$;   b) $2 \cdot 3^3 \cdot 7$;   c) $2^2 \cdot 3^2 \cdot 5 \cdot 7$.

44. Answer without calculating:

a) Is 12 a divisor of 60?   b) Is 8 a divisor of 180?   c) Is 12 a divisor of 180?

---

5. Least common multiple (LCM)

The *least common multiple (LCM)* of two or more whole numbers is the smallest natural number that is a multiple of each whole number.

The factorizations of this chapter are most often used to simplify fractions and find LCMs.

LCMs are used to compare, add, and subtract fractions. In algebra, LCMs are useful in equation solving.

You can find the LCM of two or more numbers, by using two methods:

Method 1 to calculate LCM of two or more numbers: Find the LCM of 6 and 9.

The multiples of 6 are: 6, 12, 18, 24, 30, 36, 42, 48, 54 …
The multiples of 9 are: 9, 18, 27, 36, 45, 54, 63, 72, 81 …
The common multiples of 6 and 9 are: 18, 36, 54, …
So, LCM (6, 9) = 18.

Example 4: Try yourself: Find the Least Common Multiple (LCM) of the numbers 10 and 15.

Solution: The multiples of 10 are:

The multiples of 15 are:

The common multiples of 10 and 15 are:

So the least common multiple of 10 and 15 is: LCM (10,15) =

Method 2: Using the Individual Prime-Factoring Method: Find the LCM of 12 and 15.

- Step 1: Find the prime factorization of each number in exponent form.

$$12 = 2^2 \cdot 3 \qquad 15 = 3 \cdot 5$$

- Step 2: Find the product of the highest power of each prime factor.

$$\text{LCM}(12,15) = 2^2 \cdot 3 \cdot 5 = 4 \cdot 3 \cdot 5 = \boxed{60}.$$

45. Find the LCM of 12 and 20. Do it in both ways, you should have the same result.

46. Find the LCM of 12, 18, and 30, by prime-factoring.

47. Find the LCM of:

a) 16 and 60   b) 48 and 54   c) 90 and 150   d) 40 and 90

e) 24 and 60   f) 6, 10 and 15   g) 18, 24, 40, and 60.

48. In a bus station there is a bus leaving for London every 45 minutes and one leaving for Brighton every 60 minutes. If a bus to London and a bus to Brighton leave at the same time, how many minutes will it be before two buses leave again at the same time?.

49. Mary is a doctor who works 4 days and has 1 day off. Her husband, Frank, works 5 days and has 1 day off. On May 1, they were both off. What is the next day their days off will coincide?

50. John runs 6 miles a day. John's friend runs 10 miles each day. What is the least number of days each must run so they covered the exact same distance?

51. Jose earns $240 per week. His brother earns $300 per week. What is the least number of weeks each must work so that they earn the exact same amount of money?

Exercises

The largest common factor of two or more numbers is called the **highest common factor** (HCF).

To write the highest common factor of two or more whole numbers using the Individual Prime-Factoring Method:
- Step 1: Find the prime factorization of each number in exponent form.
- Step 2: Find the product of the lowest power of **common** prime factors.

Example 5: Find the HCF of 120 and 100.

Solution: $120 = 2^3 \cdot 3 \cdot 5$ $100 = 2^2 \cdot 5^2$ $\text{LCM}(120,100) = 2^2 \cdot 5 = 4 \cdot 5 = \boxed{20}$.

52. Find the HCF of:

    a) 48 and 54   b) 90 and 150   c) 40 and 90   d) 48 and 72

    e) 24 and 60   f) 36 and 45

53. Marta has 12 red, 30 green and 42 yellow marbles and she wants to put them in boxes, as many as possible, all the boxes with the same amount of each colour and with no marbles remaining. How many boxes will she have? How many marbles of each colour are there in each box?

54. To carry 12 dogs and 18 cats, cages will be used as large as possible, and so that all fit the same number of animals. How many animals should be in each cage?

55. We want to divide a rectangular area whose dimensions are 120 m x 180 m. into square shaped plots, being as large as possible. How much should measure the side of each plot?

56. In an athletic club there are 18 boys and 24 girls. They want to make teams "unisex" as big as possible. How many members must have each team?

Exercises

1. Which of these numbers are divisible by 2?   9,   44,   50,   478,   563

2. Which of these numbers are divisible by 3?   6,   36,   63,   636,   663

3. Which of these numbers are divisible by 5?   15,   51,   255,   525,   552

4. Which of these numbers are divisible by 6?   36,   65,   144,   714

5. Which of these numbers are divisible by 9?   6,   36,   63,   636,   663

6. Which of these numbers are divisible by 10?   50,   55,   505,   550,   555

7. Which of these numbers are divisible by both 2 and 3?   444,   555,   666, 777, 888, 999

8. Which of these numbers are divisible by both 2 and 9?   450,   550,   660, 770, 880, 990

9. Which of these numbers are divisible by both 3 and 5?   445,   545,   645, 745, 845

10. Which of these numbers are divisible by both 3 and 10?   440, 550,   660, 770, 880

11. Tiger Woods finished the 4-day tournament with a stroke total of 268. Is it possible for him to have shot the same score in each of the rounds? Explain.

12. Bobbie and her five partners incurred expenses of $3060 in their satellite dish installation enterprise. Can the expenses be divided evenly in whole dollars among them? Explain.

13. Write the prime factorization of each number.

  a) 32   b) 36   c) 38   d) 42   e) 75   f) 96   g) 222   h) 276   i) 256

14. Find the LCM (Lowest Common Multiple) of each group of numbers.

  a) 20, 36   b) 36, 44   c) 24, 72   d) 25, 35   e) 28, 35   f) 40, 45

  g) 15, 50, 60   h) 32, 40, 60   i) 40, 60, 105

15. Peter has been told by his doctor to take a pill every 10 hours and a syrup every 24 hours. At this moment, he has just had the two drugs. How many hours must pass till both drug shots coincide again?

16. Three traffic lights are placed along the same avenue at three different crosses. The first one changes every 20 seconds, the second, every 30 seconds and the third every 28 seconds. They have changed to green simultaneously. How long does it take until they change again at the same time? Explain your answer.

17. Find the HCF (Highest Common Factor) of each group of numbers.

   a) 20, 36   b) 36, 44   c) 24, 72   d) 25, 35   e) 28, 35   f) 40, 45

   g) 15, 50, 60   h) 32, 40, 60   i) 40, 60, 105

18. In a bakery, they have made 240 orange cakes and 300 lemon cakes. They want to market them in bags with the same number of units without mixing the two products. How many can be put in each bag if you want them to contain the greatest number?

19. If you want to divide two strings of 40 m and 60 m into equal pieces, as large as possible, without wasting anything, how much will each piece measure?

20. Carl has a soccer game every 4 days, Matt has one every 5 days. When will they have a game on the same day again, if both had a game today?

21. A choir director of your school wants to divide the choir into smaller groups. There are 24 sopranos, 60 altos and 36 tenors. Each group will have the same number of each type of voice.

   a) What is the greatest number of groups that can be formed?

   b) How many sopranos, altos and tenors will be in each group?

22. There are 100 senators and 435 representatives in the United States of America Congress. How many identical groups could be formed from all senators and representatives (with the same number of senators and representatives in each group)?

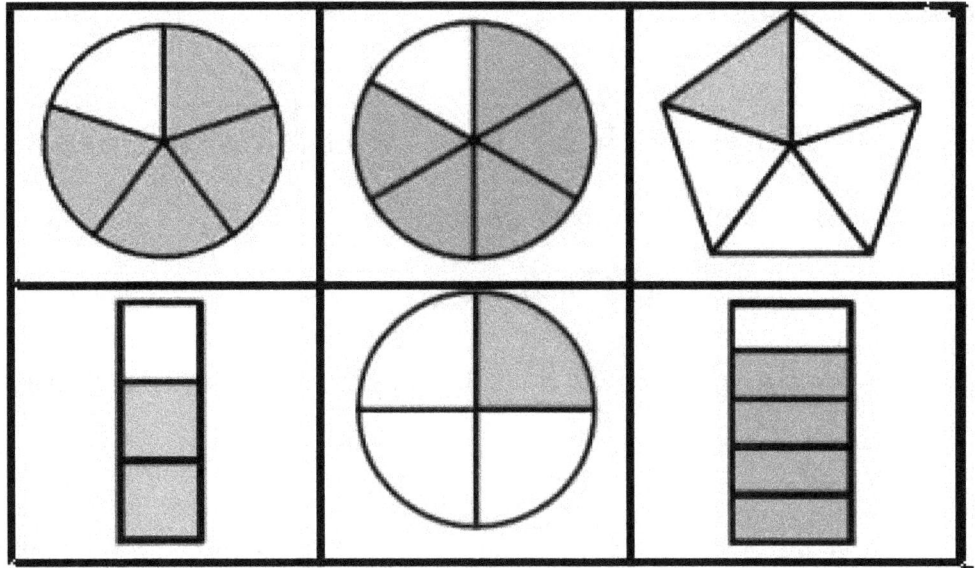

Unit 4.- Fractions

1. Concept of fraction

A *fraction* is a part of a whole.

If we cut a cake into two equal pieces, and then eat one of them, we say that we have eaten $\frac{1}{2}$ *(half)* a cake.

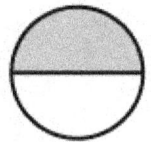

If we cut a cake into five equal pieces, then eat three of them, we say that we have eaten $\frac{3}{5}$ *(three fifths)* of a cake.

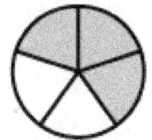

$\frac{1}{2}$ and $\frac{3}{5}$ are examples of fractions (parts of a whole).

Some definitions

- A *fraction,* (for example, $\frac{3}{5}$) is a name for a number. The upper numeral is the **numerator**. The lower numeral is the **denominator**.

- A **proper fraction** is one in which the numerator is less than the denominator. For example, $\frac{5}{12}$.

- An **improper fraction** is one in which numerator is not less than denominator. For example, $\frac{17}{12}$ or $\frac{12}{12}$.

- A **mixed number** is the sum of a whole number and a fraction with the addition sign omitted. For example, $\frac{17}{5} = 3 + \frac{2}{5}$ and it can be written as a mixed number as $3\frac{2}{5}$.

   The fraction part is usually a proper fraction.

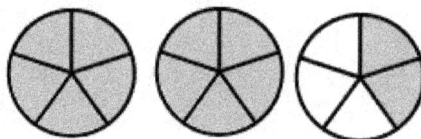

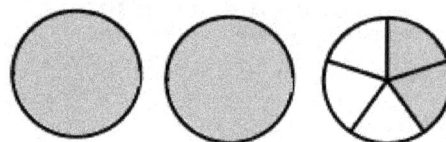

This can be seen graphically.

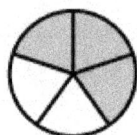

$$\frac{3}{5}$$

$$\frac{12}{5}$$

$$2\frac{2}{5}$$

Proper fraction   Improper fraction   Mixed number

2. Reading fractions

The method to name the denominator of a fraction depends on the denominator:

- If denominator ≤ 10, we use the cardinals to name it.

   For example, we read $\frac{2}{5}$ as *two fifths,* we read $\frac{9}{6}$ as *nine sixths.*

There are two exceptions, for denominators 2 and 4: $\frac{1}{2}$ is a half and $\frac{3}{4}$ is three quarters.

- For denominators larger than 10 we say "over" and do not use ordinal, so we read:

$\frac{4}{11}$ is *four over eleven* $\frac{5}{20}$ is *five over twenty* $\frac{6}{17}$ is *six over seventeen.*

1. What part of the shapes are shaded? Write in numbers and letters:

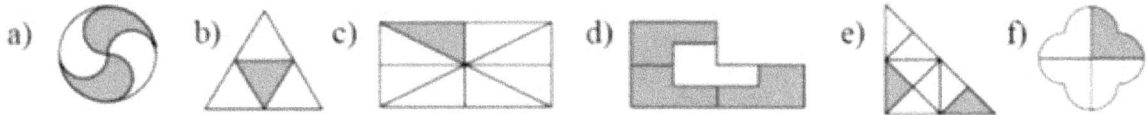

a) b) c) d) e) f)

2. Colour the given fraction of each shape:

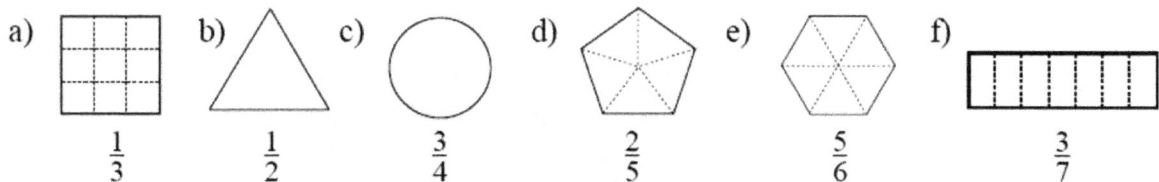

a) b) c) d) e) f)

$\frac{1}{3}$ $\frac{1}{2}$ $\frac{3}{4}$ $\frac{2}{5}$ $\frac{5}{6}$ $\frac{3}{7}$

3. In a martial arts course are 13 women and 15 men. What fraction of people do men represent and what women?

4. In a safari section of the zoo there are 7 zebras, 3 lions, 5 hippos and 1 giraffe.

a) How many animals are there together?

b) What fraction of the entries are zebras?

c) What fraction of the entries are lions?

5. Write down the way you read these fractions:

a) $\frac{5}{8}$ b) $\frac{12}{9}$ c) $\frac{5}{14}$.

6. What fraction of 1 hour is:

a) 5 minutes: b) 15 minutes: c) 40 minutes:

Imagine you are eating pizza with some friends. One of your friends cuts a pizza into 3 pieces and he eats 1 of them.

After that, you cut another pizza into 6 pieces and eat 2 of them. Who has eaten more pizza?

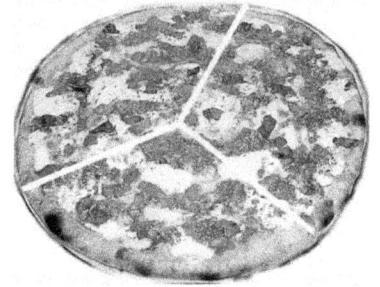

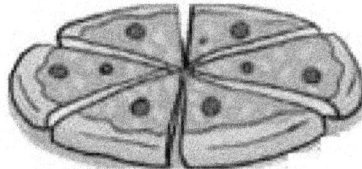

Notice you have eaten the same. These two fractions are *EQUIVALENT*

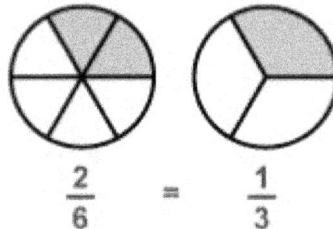

$$\frac{2}{6} = \frac{1}{3}$$

Equivalent fractions are different fractions that name the same amount.

For example:

$$\frac{2}{6} = \frac{1}{3} \qquad \frac{1}{3} = \frac{4}{12}$$

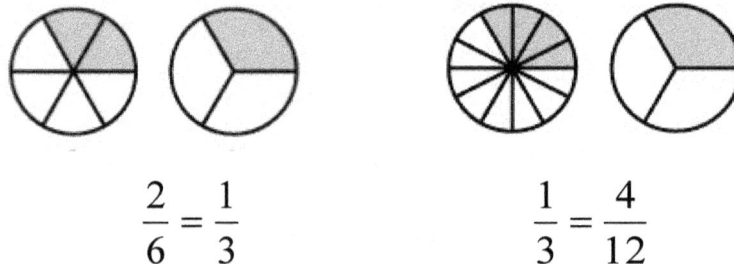

Equivalent fractions are fractions that represent equal values, even though they look different. Starting with any fractions you like, you can make up a list of equivalent fractions by simply multiplying or dividing both the numerator and the denominator by the same number each time.

7. What part of the shapes are shaded? Are they equivalent?

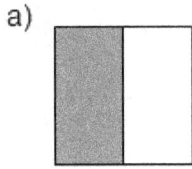

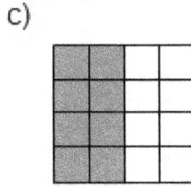

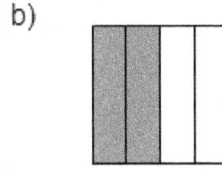

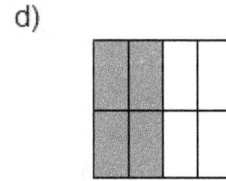

a)   c)   b)   d)

8. Copy the following figures in your notebook and colour the figures on the right to make them equivalent to the ones on the left. Write their fractions.

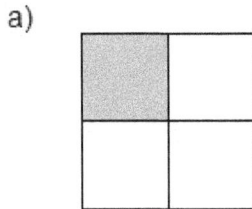

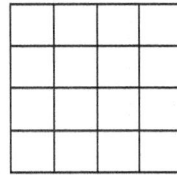

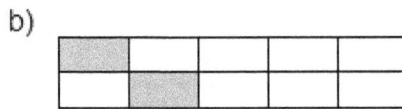

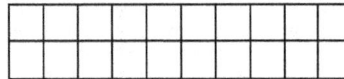

a)

b)

9. Write four equivalent fractions to the following:

a) $\dfrac{4}{6} =$ b) $\dfrac{15}{18} =$ c) $\dfrac{14}{21} =$

10. Write equivalent fractions, having the denominator given:

a) $\dfrac{2}{3} = \dfrac{}{12} = \dfrac{}{6}$ b) $\dfrac{12}{15} = \dfrac{}{5} = \dfrac{}{30}$ c) $\dfrac{7}{2} = \dfrac{}{24} = \dfrac{}{14}$

We can test if two fractions are equivalent by cross-multiplying their numerators and denominators. This is also called taking the cross-product.

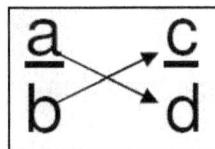

57

Example 1: Check if $\dfrac{12}{20}$ and $\dfrac{24}{40}$ are equivalent fractions.

<u>Solution:</u>

1st cross-product is the product of the first numerator and the second denominator: $12 \cdot 40 = 480$.

2nd cross-product is the product of the second numerator and the first denominator: $24 \cdot 20 = 480$.

Since the cross-products are the same, the fractions are equivalent.

Exercises

11. Test if the following pairs of fractions are equivalent:

a) $\dfrac{2}{3}$ y $\dfrac{10}{15}$ b) $\dfrac{5}{6}$ y $\dfrac{15}{12}$ c) $\dfrac{21}{15}$ y $\dfrac{84}{60}$

12. Find out the value of x that makes equivalent each pair of fractions:

a) $\dfrac{4}{6} = \dfrac{x}{9}$ b) $\dfrac{4}{10} = \dfrac{6}{x}$ c) $\dfrac{x}{21} = \dfrac{6}{9}$ d) $\dfrac{12}{20} = \dfrac{21}{x}$

4. Simplifying fractions. Simplest form of a fraction

Simplifying a fraction is the process of renaming it by using a smaller numerator and denominator.

When a fraction is *completely simplified* (its numerator and denominator have no common factors other than 1), it is s named the ***simplest form*** of the fraction.

For example $\dfrac{12}{18} = \dfrac{2}{3}$.

There are two available methods to simplify a fraction.

We must divide the numerator and denominator by a common factor. Keep dividing until there are no more common factors.

Example 2: Find the simplest form of $\dfrac{12}{30}$:

Solution: $\dfrac{12}{30} = \dfrac{12 \div 2}{30 \div 2} = \dfrac{6 \div 3}{15 \div 3} = \boxed{\dfrac{2}{5}}$

NOTE: If numerator and denominator have as last digits zeros, we can simplify all the common zeros above and below. For example, find the simplest form of $\dfrac{6000}{4500}$:

Solution: $\dfrac{6000}{4500} = \dfrac{60 \otimes \otimes}{45 \otimes \otimes} = \dfrac{60}{45} = \dfrac{60 \div 5}{45 \div 5} = \dfrac{12 \div 3}{9 \div 3} = \boxed{\dfrac{4}{3}}$

Exercises

13. Express these fractions in the simplest form. (Some may already be in the simplest form)

a) $\dfrac{5}{10}$ b) $\dfrac{30}{36}$ c) $\dfrac{18}{27}$ d) $\dfrac{48}{84}$ e) $\dfrac{45}{66}$ f) $\dfrac{64}{88}$

14. Express, in the simplest form, which fraction corresponds to these situations:

a) In a bag of 90 pens, 15 are blue

b) The number of girls and boys in our class

c) There are 90 pupils of the 270 who come by bus to the school.

Method 2: General method.

We must find the prime factorization of both numerator and denominator. Then, eliminate all common factors, other than 1, in the numerator and the denominator.

Example 3: Find the simplest form of $\dfrac{120}{100}$:

Solution: First, find the prime factorizations:

$\left.\begin{array}{l} 120 = 2 \cdot 2 \cdot 2 \cdot 3 \cdot 5 \\ 100 = 2 \cdot 2 \cdot 5 \cdot 5 \end{array}\right\}$ Then, simplify the common factors *(Do it by yourself, with your pencil!!!)*

$\left.\begin{array}{l} 120 = 2 \cdot 2 \cdot 2 \cdot 3 \cdot 5 \\ 100 = 2 \cdot 2 \cdot 5 \cdot 5 \end{array}\right\} \Rightarrow \dfrac{120}{100} = \dfrac{2 \cdot 2 \cdot 2 \cdot 3 \cdot 5}{2 \cdot 2 \cdot 5 \cdot 5} = \dfrac{2 \cdot 3}{5} = \boxed{\dfrac{6}{5}}$

NOTE: If numerator and denominator have as last digits zeros, you should simplify them at the beginning.

For example, find the simplest form of $\dfrac{7500}{5000}$:

Solution: $\dfrac{7500}{5000} = \dfrac{75 \otimes \otimes}{50 \otimes \otimes} = \dfrac{75}{50} = \left\{\begin{array}{l} 75 = 3 \cdot 5 \cdot 5 \\ 50 = 2 \cdot 5 \cdot 5 \end{array}\right\} = \dfrac{3 \cdot 5 \cdot 5}{2 \cdot 5 \cdot 5} = \boxed{\dfrac{3}{2}}$

Exercises

15. Simplify the following fractions by using the general method:

a) $\dfrac{66}{55}$ b) $\dfrac{360}{540}$ c) $\dfrac{700}{4900}$ d) $\dfrac{11 \cdot 7 \cdot 2^2 \cdot 3}{11 \cdot 2^3 \cdot 3}$

16. Simplify the following fractions by using the general method:

a) $\dfrac{126}{210}$ b) $\dfrac{126}{252}$ c) $\dfrac{2^3 \cdot 3^2 \cdot 5 \cdot 2^4}{5^2 \cdot 2^5 \cdot 3^2}$

60

Imagine you are in a party. You have eaten $\frac{2}{5}$ of a pizza and your friend Andrés has eaten $\frac{4}{7}$ of another pizza. Who has eaten more?

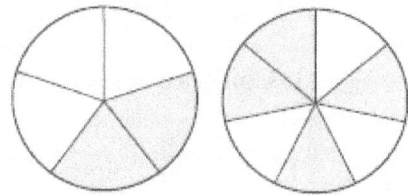

It is hard to answer this question just by looking at the fractions.

But, would it be easier if you had had $\frac{2}{5}$ of a pizza and Andrés $\frac{3}{5}$?

What is the difference?   *THEY HAVE THE SAME DENOMINATORS!!!*

So, first of all, we must transform our original fractions into two new fractions, equivalent to original ones and having the same denominator, the Lowest Common Multiple of both denominators.

This is going to be necessary also when adding or subtracting fractions, so, pay attention to this transformation into Lowest Common Denominator.

Example 4: Compare $\frac{2}{5}$ and $\frac{4}{7}$:

Solution:

Step 1: Find out the LCM(5, 7) = 35

Step 2: Transform both original fractions into two new fractions having 35 as denominator. Remember you can multiply numerator and denominator by the same number:

$\frac{2}{5} = \frac{}{35} \Rightarrow$ To transform 5 into 35, we have multiplied by 7, so, we must also multiply the numerator

by 7. So, as, $2 \cdot 7 = 14$, $\frac{2}{5} = \frac{14}{35}$

$\dfrac{4}{7} = \dfrac{}{35} \Rightarrow$ To transform 7 into 35, we have multiplied by 5, so, we also multiply the numerator

by 5. So, as, $4 \cdot 5 = 20$, $\dfrac{4}{7} = \dfrac{20}{35}$

<u>Step 3</u>: Compare the new fractions: As $\dfrac{14}{35} < \dfrac{20}{35} \Rightarrow \Rightarrow \Rightarrow \dfrac{2}{5} < \dfrac{4}{7}$

Exercises

17. Which is bigger?

    a) $\dfrac{3}{5}\, or\, \dfrac{7}{15}$ b) $\dfrac{3}{7}\, or\, \dfrac{6}{21}$ c) $\dfrac{10}{15}\, or\, \dfrac{4}{6}$ d) $\dfrac{1}{3}\, or\, \dfrac{3}{100}$

18. Put these fractions in ascending order of size: a) $\dfrac{8}{3}, \dfrac{6}{4}\, and\, \dfrac{12}{5}$ b) $\dfrac{3}{10}, \dfrac{13}{20}\, and\, \dfrac{2}{3}$

19. Insert the symbol $<, >$ or $=$:

    a) $\dfrac{2}{15}$ $\dfrac{3}{10}$ b) $\dfrac{7}{30}$ $\dfrac{5}{12}$ c) $\dfrac{3}{20}$ $\dfrac{7}{12}$ d) $\dfrac{3}{20}$ $\dfrac{2}{15}$ $\dfrac{7}{10}$

---

6. Addition and subtraction of fractions

What is the sum of $\dfrac{1}{5}$ and $\dfrac{2}{5}$?

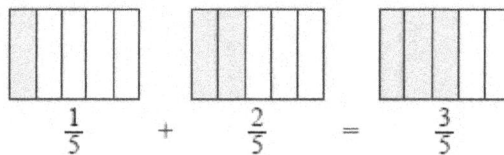

$$\dfrac{1}{5} \quad + \quad \dfrac{2}{5} \quad = \quad \dfrac{3}{5}$$

What is the difference of $\dfrac{3}{5}$ and $\dfrac{1}{5}$?

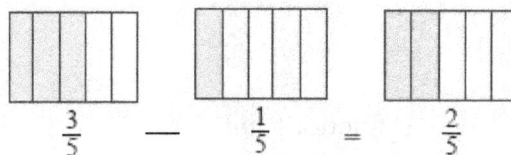

$$\dfrac{3}{5} \quad - \quad \dfrac{1}{5} \quad = \quad \dfrac{2}{5}$$

If the fractions have the same denominator the numerator of the sum is found by simply adding the numerators over the denominator. Their difference is the difference of the numerators over the denominator.

We do not add or subtract the denominators!!!

If the fractions have different denominators, it is a bit harder. Let´s see it with an example:

Example 5: Calculate $\dfrac{2}{5} + \dfrac{4}{7}$:

Solution:

Step 1: Find out the LCM(5, 7) = 35

Step 2: Transform both original fractions into two new fractions having 35 as denominator. Remember you can multiply numerator and denominator by the same number:

$\dfrac{2}{5} = \dfrac{}{35} \Rightarrow$ To transform 5 into 35, we have multiplied by 7, so, we must also multiply the numerator by 7. So, as, $2 \cdot 7 = 14$, $\dfrac{2}{5} = \dfrac{14}{35}$

$\dfrac{4}{7} = \dfrac{}{35} \Rightarrow$ To transform 7 into 35, we have multiplied by 5, so, we must also multiply the numerator by 5. So, as, $4 \cdot 5 = 20$, $\dfrac{4}{7} = \dfrac{20}{35}$

Step 3: As new fractions have the same denominator, they can be added: $\dfrac{2}{5} + \dfrac{4}{7} = \dfrac{14}{35} + \dfrac{20}{35} = \boxed{\dfrac{34}{35}}$

In this case, the final fraction cannot be simplified.

Exercises

20. Evaluate the following, expressing your answers in the simplest form.

a) $\dfrac{2}{5} + \dfrac{1}{5}$

b) $\dfrac{3}{7} + \dfrac{2}{7}$

c) $\dfrac{3}{8} + \dfrac{2}{8} + \dfrac{5}{8}$

d) $\dfrac{3}{7} - \dfrac{2}{7} + \dfrac{4}{7}$

21. Evaluate the following, expressing your answers in the simplest form.

a) $\dfrac{1}{2} + \dfrac{3}{4}$

b) $\dfrac{3}{5} + \dfrac{2}{10} + \dfrac{1}{15}$

c) $\dfrac{7}{3} + \dfrac{5}{6} - \dfrac{1}{2}$

d) $\dfrac{5}{2} + \dfrac{10}{3} - \dfrac{8}{3} - \dfrac{1}{3}$

22. Joe painted $\dfrac{2}{5}$ of a fence and Bill painted $\dfrac{1}{2}$ of it. What fraction of the fence did the boys paint?

23. In a school $\dfrac{1}{3}$ of the children eat school dinners, $\dfrac{1}{2}$ bring packed lunches and the rest go home. What fraction of the children go home for lunch?

24. Sue bought a record with $\dfrac{1}{4}$ of her allowance. She spent another $\dfrac{1}{8}$ at the cinema. What part of her allowance did she spend?

25. A group of students went to a fast food restaurant. $\dfrac{2}{5}$ of them bought a beef burger and $\dfrac{1}{3}$ of them bought a chicken burger. The rest of them just bought drinks. What fraction of the group bought drinks?

7. Product of fractions

To **multiply** two or more **fractions**, multiply the numerators and multiply the denominators, separately. Then simplify if necessary.
$$\frac{a}{b} \cdot \frac{c}{d} = \frac{a \cdot c}{b \cdot d}$$

Example 6: Calculate: a) $\dfrac{3}{8} \cdot \dfrac{5}{2}$ b) $\dfrac{5}{7} \cdot \dfrac{2}{3}$

Solution:

a) $\dfrac{3}{8} \cdot \dfrac{5}{2} = \dfrac{3 \cdot 5}{8 \cdot 2} = \dfrac{15}{16}$

b) $\dfrac{5}{7} \cdot \dfrac{2}{3} = \dfrac{5 \cdot 2}{7 \cdot 3} = \dfrac{10}{21}$

Important: If possible, you must simplify the resulting fraction. And you can do this at the end, as you already know, but you can also simplify before multiplying. Let´s see it with an example:

Example 7: Calculate $\dfrac{6}{8} \cdot \dfrac{10}{15}$

Solution: - Simplifying at the end) $\dfrac{6}{8} \cdot \dfrac{10}{15} = \dfrac{6 \cdot 10}{8 \cdot 15} = \dfrac{60}{120} = \dfrac{6}{12} = \dfrac{6 \div 2}{12 \div 2} = \dfrac{3 \div 3}{6 \div 3} = \boxed{\dfrac{1}{2}}$

- Simplifying before) $\dfrac{6}{8} \cdot \dfrac{10}{15} = \dfrac{6 \cdot 10}{8 \cdot 15} = \dfrac{(2 \cdot 3) \cdot (2 \cdot 5)}{(2 \cdot 2 \cdot 2) \cdot (3 \cdot 5)} = \dfrac{2 \cdot 3 \cdot 2 \cdot 5}{2 \cdot 2 \cdot 2 \cdot 3 \cdot 5} = \boxed{\dfrac{1}{2}}$

Note: Sometimes, it is possible to simplify even before:

Example 8: Calculate $\dfrac{6}{10} \cdot \dfrac{10}{15}$ Solution: $\dfrac{6}{10} \cdot \dfrac{10}{15} = \dfrac{6 \cdot 10}{10 \cdot 15} = \dfrac{6}{15} = \dfrac{2 \cdot 3}{3 \cdot 5} = \boxed{\dfrac{2}{5}}$

---

26. Multiply, giving your answers in the simplest form

a) $\dfrac{4}{5} \cdot \dfrac{7}{11}$ b) $\dfrac{14}{15} \cdot \dfrac{15}{19}$ c) $\dfrac{5}{4} \cdot \dfrac{2}{7}$ d) $\dfrac{5}{6} \cdot \dfrac{2}{19}$ e) $\dfrac{1}{3} \cdot \dfrac{5}{7} \cdot \dfrac{17}{12}$ f) $\dfrac{5}{9} \cdot \dfrac{2}{5} \cdot \dfrac{3}{7}$

27. Multiply, giving your answers in the simplest form:

a) $42 \cdot \dfrac{5}{6}$ b) $\dfrac{5}{4} \cdot 37$ c) $\dfrac{5}{8} \cdot 18$ d) $2 \cdot \dfrac{3}{9} \cdot \dfrac{17}{5}$ e) $\dfrac{2}{8} \cdot 3 \cdot \dfrac{3}{7}$ f) $7 \cdot \dfrac{66}{21} \cdot \dfrac{17}{13}$

28. A math's teacher spends $\dfrac{4}{5}$ of his working time in school teaching his classes. If $\dfrac{2}{9}$ of this teaching time is spent with his first year classes. What fraction of his working time does he spend teaching the first year?

29. Last night John spent $\dfrac{3}{4}$ of an hour on English homework and $\dfrac{5}{6}$ of this time, on his Maths. If he started his homework at 20 o'clock, did he finish it in time to watch The Simpsons, which started at 21 o'clock? Explain your answer.

Exercises

Calculating the fraction of a whole

Example 9: What is $\dfrac{2}{5}$ of 35?

Solution: $\dfrac{1}{5}$ of $35 = 35 \div 5 = 7$ $\Rightarrow\Rightarrow\Rightarrow$ $\dfrac{2}{5}$ of $35 = 2 \times 7 = 14$.

Example 10: How much is the $\dfrac{5}{8}$ of 32 kg?

Solution: $\quad \dfrac{1}{8}$ of $32 = 32 \div 8 = 4 \qquad \Rightarrow\Rightarrow\Rightarrow \qquad \dfrac{5}{8}$ of $32 = 5 \times 4 = 20.$

Exercises

30. Work out:

a) $\dfrac{1}{8}$ of 72 b) $\dfrac{4}{9}$ of 63 c) $\dfrac{5}{6}$ of 36 d) $\dfrac{3}{4}$ of 36

31. A chain store closed $\dfrac{2}{15}$ of its 345 shops. How many shops were closed?

32. Mum bought 1200 g of grapes. John ate 2 fifths of them, Betty ate 1 quarter of them and dad ate the rest. What amount of grapes did each of them eat?

33. Joan earns **£** 1800 a month. She spends $\dfrac{3}{8}$ of her salary every month. She gives her parents $\dfrac{2}{5}$ of the remainder and saves the rest. How much money does she save every month?

34. Mrs Holland spends $\dfrac{1}{4}$ of her money in the market and $\dfrac{1}{3}$ of the remainder in a shop. What fraction of her money is left?

35. Harban was given **£** 15 allowance each week. He spent $\dfrac{3}{5}$ of it. What fraction did he save? How much did he save in pounds?

8. Quotient of two fractions

To **divide** two fractions, turn the second fraction *UPSIDE DOWN* and then multiply them.

$$\frac{a}{b} : \frac{c}{d} = \frac{a}{b} \cdot \frac{d}{c} = \frac{a \cdot d}{b \cdot c}$$

Example 11: Calculate: $\quad \dfrac{3}{8} : \dfrac{5}{3} \qquad$ Solution: $\qquad \dfrac{3}{8} : \dfrac{5}{3} = \dfrac{3}{8} \cdot \dfrac{3}{5} = \dfrac{3 \cdot 3}{8 \cdot 5} = \dfrac{9}{40}$

Important: If possible, you must simplify the resulting fraction. And you can do this at the end, as you already know, but you can also simplify before multiplying. Let´s see it with an example:

Example 12: Calculate $\dfrac{6}{8}:\dfrac{15}{10}$

Solution:

Simplifying at the end) $\quad \dfrac{6}{8}:\dfrac{15}{10}=\dfrac{6}{8}\cdot\dfrac{10}{15}=\dfrac{6\cdot10}{8\cdot15}=\dfrac{60}{120}=\dfrac{6}{12}=\dfrac{6\div2}{12\div2}=\dfrac{3\div3}{6\div3}=\boxed{\dfrac{1}{2}}$

Simplifying before) $\dfrac{6}{8}:\dfrac{15}{10}=\dfrac{6}{8}\cdot\dfrac{10}{15}=\dfrac{6\cdot10}{8\cdot15}=\dfrac{(2\cdot3)\cdot(2\cdot5)}{(2\cdot2\cdot2)\cdot(3\cdot5)}=\dfrac{2\cdot3\cdot2\cdot5}{2\cdot2\cdot2\cdot3\cdot5}=\boxed{\dfrac{1}{2}}$

Note: Sometimes, it is possible to simplify even before:

Example 13: Calculate $\dfrac{6}{10}:\dfrac{15}{10}$

Solution: $\quad \dfrac{6}{10}:\dfrac{15}{10}=\dfrac{6}{10}\cdot\dfrac{10}{15}=\dfrac{6\cdot10}{10\cdot15}=\dfrac{6}{15}=\dfrac{2\cdot3}{3\cdot5}=\boxed{\dfrac{2}{5}}$

Exercises

36. Work out, giving your answer in the simplest form:

a) $\dfrac{4}{5}:\dfrac{7}{3}$ b) $\dfrac{2}{3}:\dfrac{1}{3}$ c) $\dfrac{10}{9}:\dfrac{5}{3}$ d) $\dfrac{5}{2}:\dfrac{1}{2}$ e) $\dfrac{7}{3}:\dfrac{2}{3}$ f) $\dfrac{8}{3}:\dfrac{6}{4}$

37. Calculate, giving your answer in the simplest form:

a) $\dfrac{3}{2}:\dfrac{4}{9}$ b) $\dfrac{1}{5}:\dfrac{5}{19}$ c) $\dfrac{3}{14}:\dfrac{4}{7}$ d) $\dfrac{15}{3}:\dfrac{5}{6}$ e) $\left(\dfrac{5}{3}:\dfrac{5}{8}\right):\dfrac{7}{4}$ f) $2:\left(\dfrac{9}{7}:\dfrac{13}{3}\right)$

38. In a factory, a 49 m long metal rod is divided into $\dfrac{7}{5}$ meter pieces. How many pieces can be made from each rod?

40. A six tons load of soil container is divided in $\dfrac{3}{140}$ tons bags. How many bags can be filled?

67

9. Order of operations

When working on fractions, the order in which the operations are performed is the same as for natural, decimal or whole numbers:

1ST: Parentheses	2ND: Exponents
3RD: Multiplication/Division	4TH: Addition/Subtraction

Exercises

41. Work out the following operations:

a) $\dfrac{3}{5} + \dfrac{7}{5} \cdot \dfrac{8}{7}$

b) $\dfrac{1}{6} + \dfrac{2}{3} \cdot \dfrac{6}{5} - \dfrac{2}{10} : \dfrac{1}{3} + \dfrac{2}{3}$

c) $\dfrac{2}{7} + \dfrac{2}{7} \cdot \dfrac{14}{3} - \dfrac{5}{3}$

d) $\dfrac{2}{13} \cdot \dfrac{26}{3} - \dfrac{2}{5} \cdot \dfrac{10}{3}$

e) $\dfrac{2}{5} - \dfrac{2}{5} \cdot \left(\dfrac{3}{4} - \dfrac{2}{6} \right) - \left(2 - \dfrac{1}{2} \right)$

Review exercises

1. Write the fraction represented by the figures.

1.

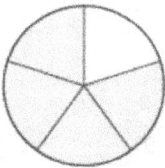

2.

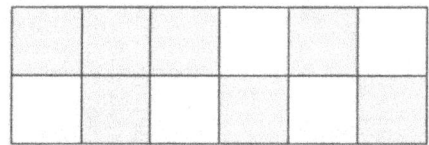

3.

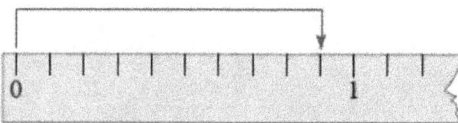

4.

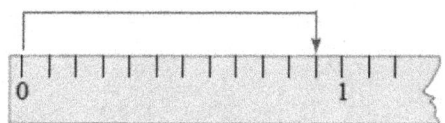

2. Identify the improper fractions from each list.

a) $\dfrac{8}{3}, \dfrac{11}{12}, \dfrac{9}{9}, \dfrac{22}{19}, \dfrac{3}{20}$

b) $\dfrac{1}{13}, \dfrac{11}{15}, \dfrac{8}{10}, \dfrac{5}{2}, \dfrac{12}{18}$

3. Change to a mixed number:

a) $\dfrac{82}{11}$　　　　b) $\dfrac{76}{9}$　　　　c) $\dfrac{41}{5}$

4. Change to an improper fraction:

a) $6\dfrac{5}{12}$　　　　b) $4\dfrac{3}{7}$　　　　c) $9\dfrac{2}{3}$

5. Write four fractions, equivalent to each of the given fractions:

a) $\dfrac{2}{3}$　　　　b) $\dfrac{2}{7}$　　　　c) $\dfrac{4}{11}$

6. Find the missing numerator:

a) $\dfrac{4}{5} = \dfrac{?}{45}$　　　　b) $\dfrac{6}{7} = \dfrac{?}{56}$　　　　c) $\dfrac{5}{6} = \dfrac{?}{144}$

7. List the fractions from smallest to largest:

a) $\dfrac{3}{5} , \dfrac{2}{5} , \dfrac{1}{4} , \dfrac{1}{7}$　　　　b) $\dfrac{2}{9} , \dfrac{3}{5} , \dfrac{6}{15}$　　　　c) $\dfrac{6}{8} , \dfrac{5}{4} , \dfrac{5}{6} , \dfrac{10}{8}$

8. Simplify:

a) $\dfrac{24}{32}$　　b) $\dfrac{60}{90}$　　c) $\dfrac{35}{70}$　　a) $\dfrac{102}{6}$　　b) $\dfrac{126}{42}$　　c) $\dfrac{75}{125}$

9. My sister´s child is 219 days old. What fraction of a year (365 days) old is she?

10. On a math test, a student answers 42 items correctly and 18 incorrectly. What fraction of the items are answered correctly? Simplify.

11. Add. Simplify completely:

a) $\dfrac{7}{30} + \dfrac{7}{30} + \dfrac{6}{30}$　　　　b) $\dfrac{4}{15} + \dfrac{1}{3}$　　c) $\dfrac{3}{35} + \dfrac{8}{21}$　　　　d) $\dfrac{11}{30} + \dfrac{9}{20} + \dfrac{3}{10}$

12. In order to complete a project, Preston needs $\dfrac{3}{10}$ cm of foam, $\dfrac{2}{5}$ cm of metal $\dfrac{2}{5}$ cm of wood and $\dfrac{11}{15}$ cm of plexiglass. What will be the total thickness of this project when these materials are piled up?

13. Subtract:

a) $\dfrac{7}{8} - \dfrac{5}{16}$　　　　b) $\dfrac{17}{24} - \dfrac{1}{6}$　　　　c) $\dfrac{7}{15} - \dfrac{3}{20}$　　　　d) $\dfrac{7}{10} - \dfrac{1}{4}$

14. Wanda finds $\frac{3}{4}$ oz of gold during a day of panning along the Snake River. She gives a $\frac{1}{3}$ oz nugget to Jose, her guide. What fraction of an ounce of gold does she have left?

15. Dolly opened a $\frac{3}{4}$ liter bottle of coke, James drank $\frac{1}{3}$ liter and Carole $\frac{1}{5}$ liter. How much coke was left for Dolly?

16. Calculate and simplify:

a) $\dfrac{6}{10} \cdot \dfrac{5}{3}$

b) $\dfrac{8}{7} \cdot \dfrac{14}{24} \cdot \dfrac{15}{5}$

c) $\dfrac{2}{9} \cdot \dfrac{18}{6} \cdot \dfrac{3}{2}$

d) $3 \cdot \dfrac{5}{2}$

e) $7 \cdot \dfrac{3}{14}$

f) $4 \cdot \dfrac{1}{6} \cdot \dfrac{3}{2}$

g) $11 \cdot \dfrac{3}{22} \cdot \dfrac{1}{6}$

h) $\dfrac{6}{10} : \dfrac{5}{3}$

i) $\dfrac{2}{5} : \dfrac{4}{3}$

17. Work out the following operations:

a) $\dfrac{3}{5} + \dfrac{3}{5} \cdot \dfrac{1}{3} - \dfrac{3}{10} : \dfrac{1}{4}$

b) $\left(\dfrac{2}{7} - 2 \right) \cdot \left(1 - \dfrac{5}{4} - \dfrac{25}{12} \right)$

c) $\left(\dfrac{3}{5} - \dfrac{1}{2} \right) : \dfrac{3}{10}$

Unit 5.- Decimal numbers

1. Decimal numbers. Reading decimal numbers

Decimal numbers, or simply, *decimals,* such as 3.762, are used in situations in which we look for more precision than whole numbers provide.

As with whole numbers, a digit in a decimal number has a value which depends on the place of the digit. In our decimal system, digits can be placed to the left and right of a decimal **point**, to indicate numbers greater than one or less than one, respectively. The decimal point helps us to keep track of where the "ones" place is. It's placed just to the right of the ones place. As we move right from the decimal point, each number place is divided by 10.

Note that adding extra zeros to the right of the last decimal digit does not change the value of the decimal number. This means that 123.456 = …..0000123.4560000……..

71

Place (underlined)	Name of Position
<u>1</u>.234567	Ones (units) position
1.<u>2</u>34567	Tenths
1.2<u>3</u>4567	Hundredths
1.23<u>4</u>567	Thousandths
1.234<u>5</u>67	Ten thousandths
1.2345<u>6</u>7	Hundred Thousandths
1.23456<u>7</u>	Millionths

	Whole-number part				Decimal point	Fractional part				
Ten thousands	Thousands	Hundreds	Tens	Ones		Tenths	Hundredths	Thousandths	Ten-thousandths	Hundred-thousandths
··· 10,000	1000	100	10	1		$\frac{1}{10}$	$\frac{1}{100}$	$\frac{1}{1000}$	$\frac{1}{10,000}$	$\frac{1}{100,000}$ ···
··· 10^4	10^3	10^2	10^1	1		$\frac{1}{10^1}$	$\frac{1}{10^2}$	$\frac{1}{10^3}$	$\frac{1}{10^4}$	$\frac{1}{10^5}$ ···

How do we write the name of a decimal number?

To write the name for a decimal number

1. Write the name for the whole number to the left of the decimal point.

2. Write the word *and* for the decimal point.

3. Write the whole number name for the number to the right of the decimal point.

4. Write the place value for the digit farthest to the right.

If the decimal has only zero or no digit to the left of the decimal point, omit steps 1 and 2.

Example 1:

0.83	is read as	Eighty-three hundredths
0.0027	is read as	Twenty-seven ten-thousandths
556.43	is read as	Five hundred fifty-six and forty-three hundredths

---

To write the place value name for a decimal

1. Write the whole number (number before the word *and*).

2. Write a decimal point for the word *and.*.

3. Ignoring the place value name, write the name for the number following the word *and*. Insert zeros if necessary, between the decimal point and the digits following it to ensure that the place on the far right has the correct (given) place value.

Example 2:

- Two hundred thirty-seven and fifty-eight hundredths   is written   237.58

- Seven hundred twenty-three thousandths   is written   0.723

---

1. Write in words the following decimals:

a) 0.26   b) 0.82   c) 0.267   d) 0.943   e) 7.002   f) 4.3007

g) 11.92   h) 32.03   i) 0.805   j) 8.05   k) 80.05   l) 8.0 05

m) 61.0203   n) 45.0094   ñ) 90.003   o) 900.030   p) 21.456   q) 0.77

r) 0.0089   s) 5.7254

2. Write the place value name.

a) Forty-two hundredths   b) Sixty-nine hundredths

c) Four hundred nine thousandths   d) Five hundred nineteen thousandths

e) Nine and fifty-nine thousandths   f) Sixteen and six hundredths

g) Three hundred eight ten-thousandths   h) Twelve ten-thousandths

Exercises

How much would you say it costs??

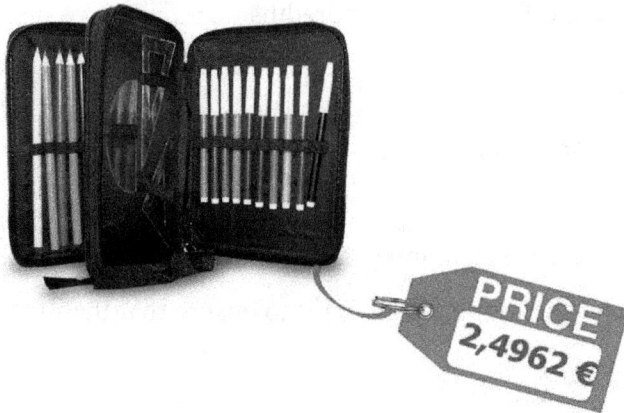

It costs, aproximately, **2.5 €**

To round a decimal number to a given place value

1. Draw an arrow under the given pace value. (After enough practice, you will be able to round mentally and will not need the arrow).

2. If the digit to the right of the arrow is 5, 6, 7, 8 or 9, add 1 to the digit above the arrow; that is, round to the larger number.

3. If the digit to the right of the arrow is 0, 1, 2, 3 or 4, keep the digit above the arrow; that is, round to the smaller number.

4. Write whatever zeros are necessary after the arrow so that the number above the arrow has the same place value as the original.

Example 3:

a) Round 0.7539 to the nearest hundredth.

$0.7539 \approx 0.75$ The digit to the right of the round-off place is 3, so round
    ↑   down.

b) Round 7843.9 to the nearest thousand.

7843.9 ≈ 8000
↑   Three zeros must be written after the 8 to keep it in the thousands place.

c) Round 537.7 to the nearest unit.

537.7 ≈ 538
↑   The number to the right of the round-off place is 7, so we round up by adding 537 + 1 + 538.

3. Round to the nearest hundredth:

   a) 67.4856   b) 27.6372   c) 548.7235   d) 375.7545

4. Build a table and, in it, round to the nearest ten, hundredth and thousandth:

   a) 35.7834   b) 61.9639   c) 86.3278   d) 212.7364

   e) 0.91486   f) 0.8049

5. Round to the nearest dollar:

   a) $72.49   b) $38.51   c) $7821.51   d) $8467.80

3. Comparing decimals

Symbols < (less or lower than), > (greater or bigger than) and = (equals) are used to show how the size of one number compares to another.

To list a set of decimals from smallest to largest

1. Make sure that all numbers have the same number of decimal places to the right of the decimal point by writing zeros to the right of the last digit when necessary.

2. Write the numbers in order as if they were whole numbers.

3. Remove the extra zeros.

Example 4: List these numbers from smallest to largest: 4.502   4.511   4.6005   4.50102.

Solution: As you can see, the number that has more decimal digits is 4.50102, with 5 of them. So, we add zeros to the others, so that they all have 5 decimal digits → 4.50200; 4.51100; 4.60050; 4.50102.

If they were whole numbers, they would be → 450200; 451100; 460050; 450102.

Now, we order them as whole numbers → 450102 < 450200 < 451100 < 460050.

Finally, we substitute them by originals (removing added zeros) →

4.50102 < 4.502 < 4.511 < 4.6005.

---

6. Arrange these numbers in order of size, smallest first:

6.21,   6.023,   6.4,   6.04,   2.71,   9.4

7. Insert the symbols < or > between these pairs of numbers

a) 1.2__0.62   b) 1.23__1.3   c) 4.008__4.03

d) 0.24__0.204   e) 0.509__0.6   f) 1.582__1.59

8. Arrange these numbers in order of size, smallest first:

a) 6.1; 4.22; 4.02; 6.11; 3.99; 3.9   b) 5.602; 5.611; 5.6005; 5.60102

9. Arrange these numbers in order of size, smallest first:

a) 0.75; 0.57; 0.507; 0.705   b) 0.102; 0.05; 0.105; 0.501; 0.251

10. Insert a decimal number in each pair:

a) 4.5 and 4.6   b) 7.24 and 7.242   c) 2.3 and 2.4   d) 3.35 and 3.36

e) 5.23 and 5.24   f) 5.39 and 5.4   g) 3 and 3.1   h) 6.03 and 6.04

11. Is the statement true or false?

a) 0.38 < 0.3   b) 0.49 < 0.50   c) 10.48 > 10.84   d) 7.78 < 7.87

Exercises

To add or subtract decimals, line up the decimal points and then follow the rules for adding or subtracting whole numbers, placing the decimal point in the same column as above.

When one number has more decimal places than another, use zeros to give them the same number of decimal places.

To add decimal numbers

1. Write in columns with the decimal points aligned. Insert extra zeros to help align the place values.

2. Add the decimals as if they were whole numbers.

3. Align the decimal point in the sum with those above.

Example 5:   Add $43.67 + 2.3$ Subtract $57.8 - 8.06$

$$\begin{array}{r} a) \quad 43.67 \\ + \ 2.30 \\ \hline 45.97 \end{array} \qquad \begin{array}{r} b) \quad 57.80 \\ - \ 8.06 \\ \hline 49.74 \end{array}$$

Exercises

12. Add:

   a) $0.7 + 0.7$ b) $0.6 + 0.5$ c) $3.7 + 2.2$ d) $7.6 + 2.9$

   e) $1.6 + 5.5 + 8.7$ f) $6.7 + 2.3 + 4.6$ g) $34.8 + 5.29$ h) $22.9 + 7.67$

13. The sum of 6.7, 10.56, 5.993, and 45.72 has _____ decimal places.

14. Add:

   a) $2.337 + 0.672 + 4.056$ b) $9.445 + 5.772 + 0.822$

   c) $0.0017 + 1.007 + 7 + 1.071$ d) $1.0304 + 1.4003 + 1.34 + 0.403$

15. Subtract:

a) 0.831 - 0.462

b) 0.067 - 0.049

c) 33.456 - 29.457

d) 7.598 - 4.7732

e) 327.58 - 245.674

f) 506.5065 - 341.341

16. On a vacation trip, Manuel stopped for gas four times. The first time, he bought 19.2 gallons. At the second station he bought 21.9 gallons, and at the third, he bought 20.4 gallons. At the last stop, he bought 23.7 gallons. How much gas did he buy on the trip?

17. What is the total cost of a cart of groceries that contains bread for $3.09, bananas for $1.49, cheese for $2.50, cereal for $4.39, coffee for $7.99, and meat for $9.27? Round the result to the tenths.

18. The table shows the lengths of railway tunnels in various countries.

World's Longest Railway Tunnels

Tunnel	Length (km)	Country
Seikan	53.91	Japan
English Channel Tunnel	49.95	UK–France
Dai-shimizu	22.53	Japan

a) How much longer is the longest tunnel than the second longest tunnel?

b) What is the total length of the Japanese tunnels?

19. In 2004, the average interest rate on a 30-year home mortgage was 6.159%. In 2009, the average interest rate was 4.759%. What was the drop in interest rate?

20. How high from the ground level is the top of the tree shown below? Round to the nearest foot.

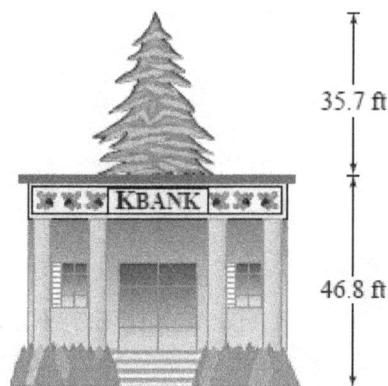

35.7 ft

46.8 ft

KBANK

78

5. Multiplying decimal numbers

Multiplying decimals is just like multiplying whole numbers. The only extra step is to decide how many digits to leave to the right of the decimal point. To do that, add the numbers of digits to the right of the decimal point in both factors.

To multiply decimals
1. Multiply the numbers as if they were whole numbers.
2. Locate the decimal point by counting the number of decimal places (to the right of the decimal point) in both factors. The total of these two counts is the number of decimal places the product must have.
3. If necessary, zeros are inserted to the left of the numeral so there are enough decimal places.

Example 6: Multiply 23.56 x 34.1

Solution:

$$
\begin{array}{r}
23.56 \\
\times\ 3.1 \\
\hline
2356 \\
7068\ \ \ \\
\hline
73.036
\end{array}
$$

Decimal digits: **3**

79

23. Calculate:

   a) 5.6 x 6.9   b) 12.37 x 76.78   c) 4.66 · 4.7   d) 0.345 · (32.4 − 4.67)

24. Find the product of 9.73 and 6.8.

25. Multiply 7.9 times 0.0004.

26. Multiply 32 x 0.846 and round the product to the nearest tenth.

27. The table shows calories expended for some physical activities.

Calorie Expenditure for Selected Physical Activities

Activity	Step Aerobics	Running (7 min/mile)	Cycling (10 mph)	Walking (4.5 mph)
Calories per pound of body weight per minute	0.070	0.102	0.050	0.045

a) Vanessa weighs 145 lb and does 75 min of step aerobics per week. How many calories per week does she burn per week?

b) Steve weighs 187 lb and runs 25 min five times at a 7 min/mi pace. How many calories does he burn per week?

28. An order of 43 bars of steel is delivered to a machine shop. Each bar is 17.325 ft long. Find the total linear feet of steel in the order.

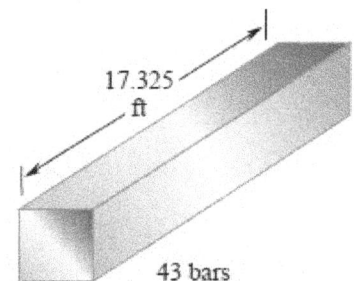

17.325 ft

43 bars

29. In the 2009 World Championships in Athletics, Shelly-Ann Fraser of Jamaica won the 100 m with a time of 10.73 sec. If she could continue that rate, what would her time for the 400 m be?

30. In the 2009 World Championships in Athletics, Usain Bolt set a new world record for 200 m, with a time of 19.19 sec. Assuming he ran the first 100 m in his record time of 9.58 sec, how long did the second 100 m take him?

6. Multiplying and dividing by a power of ten

Sometimes, e.g. when dividing two decimals, it is necessary to multiply a decimal number by a power of ten. But it will be more interesting the inverse process, converting a decimal number containing a large quantity of zeros into the product of a decimal and a power of ten.

To **multiply** a decimal by a power of ten, we must to move the decimal point to the right. The number of places to move is shown by the exponent in the power of 10.

To **divide** a decimal by a power of ten, we must to move the decimal point to the left. The number of places to move is shown by the exponent in the power of 10.

In both cases, we can add zeros if necessary.

* **IMPORTANT:** If the exponent of the power of ten is negative, the movements have an inverse direction.

Example 7: Calculate: a) $34.7561 \cdot 10^2$ b) $12345.8 \cdot 10^{-3}$ c) $12.5878 : 10^2$ d) $3.4 : 10^{-3}$.

Solution:

a) $34.7561 \cdot 10^2 = 3475.61$ b) $12345.8 \cdot 10^{-3} = 12.3458$

c) $12.5878 : 10^2 = 0.125878$ d) $3.4 : 10^{-3} = 3400$

31. Calculate the following products:

    a) $34.7 \cdot 10^2$ b) $0.892 \cdot 10$ c) $83 \cdot 10^{-3}$ d) $4.5 \cdot 10^4$ e) $23 \cdot 10^{-2}$

    f) $45.2 \cdot 10^{-3}$ g) $0.98 \cdot 10^{-1}$

32. Calculate the following quotients:

    a) $34.7 : 10^2$ b) $0.892 : 10^{-1}$ c) $83 : 10^3$ d) $4502.1 : 10^4$ e) $7.41 : 10^6$

    f) $231.6 : 0'1$ g) $583 : 10^{-4}$ h) $2.14 : 10^3$ i) $3901.3 : 10^{-3}$

33. Calculate the following products and quotients:

    a) $34.12 \cdot 10^3$ b) $2120.8 \cdot 10^{-5}$ c) $28.1 : 10^2$ d) $432 \cdot 10^{-4}$

    e) $5.786 \cdot 10^5$ f) $2.45 : 10^3$ g) $8.123 : 10^{-2}$ h) $456.1 : 10^{-1}$

7. Dividing decimal numbers

7.1. Dividing whole numbers, with decimals

Sometimes, when dividing two whole numbers, if the division is not exact, we can calculate the whole quotient and remainder, or we can also continue the whole division adding zeros to the right of the dividend until we get the amount of decimal digits required.

Example 8: Divide 235:6 until the hundredth:

```
235|00 |6
 55|    39.16
  1|0
   |40
    4//
```

34. Calculate with two decimal digits:

   a) 56 : 7   b) 7634 : 34   c) 679 : 32   d) 9783 : 127

7.2. Dividing decimals by decimals

To divide by a decimal,

1. Multiply that decimal by a power of 10 great enough to obtain a whole number.

2. Multiply the dividend by that same power of 10.

3. Now the problem becomes one involving division by a whole number instead of division by a decimal.

Example 9: Divide 0.35789 : 0.12 until the hundredths:

<u>Solution:</u>   First of all, we must multiply $0.12 \times 10^2 = 12$,   and also $0.35789 \times 10^2 = 35.789$

35. Calculate with three decimal digits:

   a) 56.7 : 2.34   b) 1432.3 : 0.42   c) 12.34 : 3.5   d) 1 : 1.2

36. 17 tickets cost £ 21.25. If they all cost the same, find the cost of one ticket.

37. A bottle contains 0.9 liters of lemonade. How many glasses with a capacity of 0.15 liters, can be filled from it?

38. A milkman is carrying a crate which contains 12 bottles and weighs 11.5 kg. If the empty crate weighs 0.7 kg, what is the weight of each bottle of milk?

Decimal expressions of rational numbers

As a fraction is the indicated quotient of two numbers. we can calculate these quotients only by dividing numerator and denominator.

For example. if we divide numerator by denominator of the fractions $\frac{221}{4}$. $\frac{25}{3}$ and $\frac{59}{6}$. we will obtain their *decimal expressions*:

$$\frac{221}{4} = 55.25 \qquad \frac{25}{3} = 8.3333....... \qquad \frac{59}{6} = 9.833333.....$$

As you can see. when dividing numerator by denominator will happen on of these situations:

> ➢ Division finishes because it is exact. as happens in the first fraction. In this case. we obtain a **limited** decimal expression (it has a limited number of decimal digits). Its decimal expression is said to be an *exact or terminating decimal number*.

> ➢ Division does not finish because it is not exact. as happens in second and third fractions. In these cases. we obtain a **non-limited** decimal expression (its decimal digits do not finish). Their decimal expressions are named *recurring* or *repeating decimal numbers* Second one is named *pure recurring decimal number*. and third one is a mixed recurring decimal number.

Repeating part is named *period* and it is indicated with an arch on it. For example:

$$2.333........ = 2.\overline{3} \qquad\qquad 1.8333........ = 1.8\overline{3}$$

So. decimal numbers can be classified into:

> *Exact* or *terminating* decimal: It finishes. so you can write down all its digits.

  For example: 0.125

> *Recurring* or *repeating* decimal: It does not finish. it goes on forever. but <u>some of the digits are repeated over and over again.</u>

  For example: $5.6666....... = 5.\overline{6}$ $14.353535....... = 14.\overline{35}$ $87.42222...... = 87.4\overline{2}$

  Between recurring decimals. we can distinguish:

> *Pure recurring decimals*: Recurring decimals in which period starts just after the decimal point.

  For example: $5.6666...... = 5.\overline{6}$ $14.353535....... = 14.\overline{35}$

> *Mixed recurring decimals:* Recurring decimals in which period does not start just after The decimal point.   For example: $87.42222...... = 87.4\overline{2}$

Decimal fractions

Fractions whose decimal expression is a terminating decimal are named *decimal fractions.*

There is a very simple rule to identify decimal fractions without calculating their decimal expression:

A fraction is a *decimal fraction* if its denominator has as prime factors. only *2's* and/or *5's*.

Exercises

39. Express in its decimal form the following fractions and indicate their type of decimal number:

a) $\dfrac{9}{2}$ b) $\dfrac{12}{7}$ c) $\dfrac{4}{3}$ d) $\dfrac{17}{100}$ e) $\dfrac{1}{6}$ f) $\dfrac{7}{495}$ g) $\dfrac{25}{99}$

40. Copy the following fractions in your notebook and round decimal fractions. You do not need to calculate them.

a) $\dfrac{19}{52}$ b) $\dfrac{41}{18}$ c) $\dfrac{37}{75}$ d) $\dfrac{17}{125}$ e) $\dfrac{1}{40}$ f) $\dfrac{7}{300}$ g) $\dfrac{8}{99}$

Rational expression of decimal numbers

All decimal numbers can be expressed as a fraction.

In following example we show the process to find out these fractions.

· **Exact decimal:** Express as a fraction $x = 2.34$

<u>Solution:</u> We must build a fraction whose

- Numerator: Decimal number without decimal point.

- Denominator: A "1" followed by as many "0" as decimal digits decimal number has.

In our case: $2.34 = \dfrac{234}{100}$ If possible. we must simplify this fraction to give it in its simplest

form $2.34 = \dfrac{224}{100} = \dfrac{112}{50} = \boxed{\dfrac{56}{25}}$

· **Recurring decimal number:** Express as a fraction $x = 14.\overline{35} = 14.3535355.........$

<u>Solution:</u> We must apply the following process. consisting on multiplying it by 1. 10. 100. 1000. ..... till finding two decimals having *identical decimal parts*. so that they can be subtracted. to cancel their decimal parts and converting them into integers.

$$100 \cdot x = 1435.353535....$$
$$10 \cdot x = 143.535353.....$$
$$x = 14.353535........$$

As you can see. x and 100·x have the same decimal part. So. we subtract them:

$$100 \cdot x = 1435.353535....$$
$$\cancel{10 \cdot x = 143.535353.....}$$
$$x = 14.353535........$$

$$99 \cdot x = 1421 \implies x = \frac{1421}{99}$$

Example 5: Express as a fraction: a) $8.\overline{5}$ b) $25.4\overline{52}$

<u>Solution:</u>

a)
$$10 \cdot x = 85.555.......$$
$$x = 8.5555......$$

As you can see. both decimals have the same decimal part. So. we subtract them:

$$10 \cdot x = 85.555.......$$
$$x = 8.5555......$$

$$9 \cdot x = 77 \implies x = \frac{77}{9}$$

b)

$$1000 \cdot x = 25452.5252......$$
$$100 \cdot x = 2545.25252......$$
$$10 \cdot x = 254.525252....$$
$$x = 25.4525252....$$

As you can see. $1000 \cdot x$ and $10 \cdot x$ have the same decimal part. So. we subtract them:

$$1000 \cdot x = 25452.5252......$$
$$\cancel{100 \cdot x = 2545.25252......}$$
$$\cancel{10 \cdot x = 254.525252....}$$
$$x = 25.4525252....$$

$$990 \cdot x = 25198 \implies x = \frac{25198}{990}$$

41. Write each of the following *exact* decimals as a fraction:

   a) 23.45   b) 5.101   **Sol.:** a) 469/20 ; b) 5101/1000.

42. Write each of the following *pure recurring* decimals as a fraction:

   a) $24.\overline{3}$ b) $0.\overline{678}$ c) $32.\overline{76}$ d) $1.\overline{876}$ e) $26.\overline{7}$

43. Write each of the following *mixed recurring* decimals as a fraction:

   a) $2.7\overline{5}$ b) $34.6\overline{41}$ c) $2.09\overline{5}$ d) $0.2\overline{3}$ e) $5.1\overline{204}$

   f) $103.25\overline{6}$ g) $0.0\overline{23}$

44. Write each of the following decimals as a fraction:

   a) $1,\overline{04} =$ b) $7,\overline{123} =$ c) $8,1\overline{74} =$ d) $1,7\overline{6} =$ e) $2,5\overline{19} =$

45. Write each of the following decimals as a fraction:

   a) $23.\overline{45}$ b) $76.8\overline{1}$ c) 398.12   d) 5.6   e) $28.\overline{7}$

   f) $261.1\overline{23}$ g) $0.0\overline{4}$ h) 23.05

87

Review exercises

1. Write the word name.

    a) 6.12   b) 0.843   c) 15.058   d) 0.0000027

2. Write the place value name.

  a) Twenty-one and five hundredths   b) Four hundred nine ten-thousandths

  c) Four hundred and four hundredths   d) One hundred twenty-five and forty-five thousandths.

3. Round the numbers to the nearest tenth, hundredth, and thousandth.

	Tenth	Hundredth	Thousandth
9. 34.7648			
10. 7.8736			
11. 0.467215			

4. The display on Mary's calculator shows 91.457919 as the result of a division exercise. If she is to round the answer to the nearest thousandth, what answer does she report?

5. List the set of decimals from smallest to largest.

    a) 0.95, 0.89, 1.01   b) 0.09, 0.093, 0.0899

    c) 7.017, 7.022, 0.717, 7.108   d) 34.023, 34.103, 34.0204, 34.0239

6. Is the statement true or false?

    a) 6.1774 > 6.1780   b) 87.0309 < 87.0319

7. Find the sum of 3.405, 8.12, 0.0098, 0.3456, 11.3, and 24.9345.

8. Find the difference of 56.7083 and 21.6249.

9. Find the perimeter of the following figure:

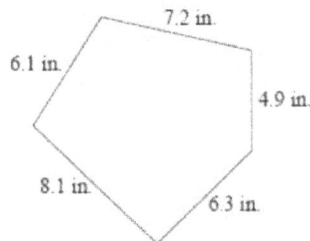

10. Multiply: 0.074 x 2.004. Round to the nearest thousandth.

11. Multiply: 0.0098 x 42.7. Round to the nearest hundredth.

12. Find the area of the rectangle.

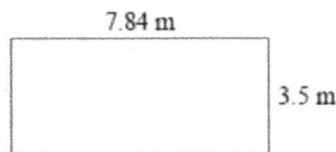

13. Multiply or divide:

    a) $13.765 \cdot 10^3$ b) $7.023 \cdot 10^6$ c) $0.7321 \cdot 10^5$ d) $9.503 \cdot 10^2$

14. Multiply or divide:

    a) $7 \cdot 10^7$ b) $8.13 \cdot 10^{-6}$ c) $6.41 \cdot 10^{-2}$ d) $3.505 \cdot 10^3$

15. Divide, rounding to the nearest tenth:

    a) 0.3 : 0.0111   b) 75 : 40.5   c) 56.7 : 0.32

16. Carol drove 375.9 km on 12.8 gallons liters of fuel. How many kilometers does she drive on each liter? Calculate with two decimal digits.

17. The Metropolis Police Department reported the following number of robberies for the week:

    Monday: 12   Tuesday: 21   Wednesday: 5   Thursday: 18

    Friday: 46   Saturday: 67   Sunday: 17.

    To the nearest tenth, what is the average number of robberies reported per day?

18. Perform the indicated operations:

    a) 0.65 + 4.29 - 2.71 + 3.04   b) $13.8 : 0.12 \cdot 4.03$ c) $(6.7)^2 - 4.4 \cdot 2.93$
    d) $2.4 \cdot (3.02 + 0.456) - (9.231 + 0.4)$ e) $100.15 : 100 - 3.995 \cdot 0.05$

19. Ellen earns £137.40 per week and after 4 weeks she gets an extra payment of £24.75. She spends £354.60 in this period. How much has she saved?

20. Susan cooked a cake and used 1.35 kg of flour, 0,37 kg of sugar, 3 eggs that weigh 80 g each and 240 g of milk. Which is the weight of the mixture?

21. I bought 7 mugs and a glass and paid 58 €. The price of the glass was 4.45€ . How much did each mug cost?

22. Express as a fraction:

   a) 2.41   b) 0.75   c) $3.\overline{5}$ d) $1.0\overline{3}$

23. Express as a fraction and check that the result of these operations is an integer:

   a) 2.333... + 4.666.....   b) 6.171717... + 3.828282....

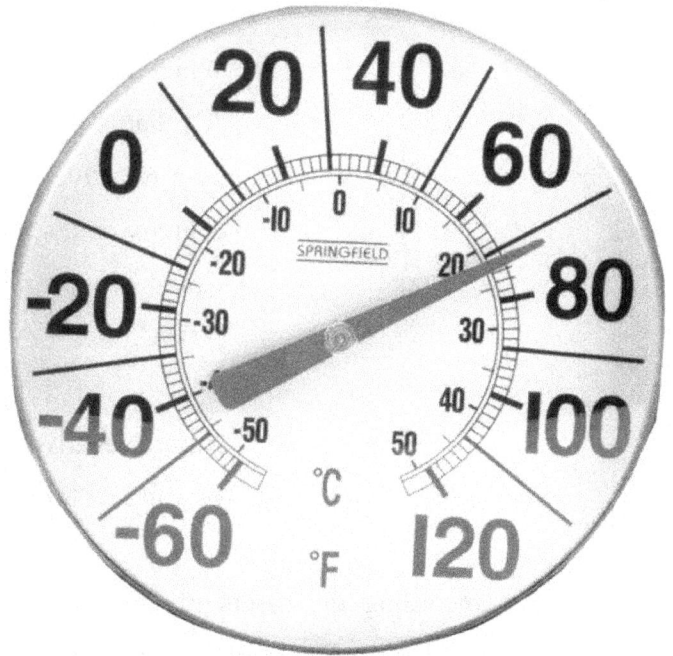

Unit 6.- Integers

1. Negative numbers

Negative numbers are an extremely useful tool for many kinds of problems. For instance, the concept of a negative temperature, the notion of a negative balance in a bank account, and the use of negative numbers to describe the motion of an object in a particular direction are important uses of negative numbers.

The following examples show specific situations where negative numbers are used.

On the Celsius temperature scale, the freezing point of water is 0°C. When we want to represent a temperature in Celsius degrees (written °C) that is colder (less than) the freezing point of water, we represent it by a negative number of degrees. For instance, a temperature that is 7 degrees colder than the freezing point of water on the Celsius scale would be represented by using the negative number (-7), and written as (-7) °C.

Note: Negative numbers are usually written into parentheses. This is very useful for us to be alert when they appear.

When recording the assets of corporations, a deficit (a situation where the corporation owes more money than it has assets) can be described by saying that the company has negative assets. So, for instance, if the corporation has $3,000,000 in assets but owes $4,000,000, then we say the corporation has assets of "negative one million dollars," written as -$1,000,000.

When measuring changes in a quantity, such as a change in temperature or a change in rainfall from one year to the next, we use a negative number to describe a change in which the quantity decreases. For instance, if the temperature at noon is 20°C and the temperature at 6 P.M. is 12°C, we say that the change in temperature from noon to 6 P.M. is -8°C.

Historically, the recognition and use of negative numbers developed very late. Despite the fact that mathematicians in ancient civilizations did perform subtraction, the recognition of negative numbers as "legitimate" numbers did not occur until the 1600s.

The Number Line

The number line is a line labeled with the integers in increasing order from left to right, that extends in both directions:

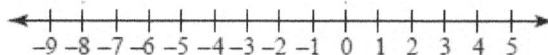

For any two different places on the number line, the integer on the right is greater than the integer on the left.

For example, $9 > 4$ Is read: nine is **'greater than'** four

 $(-7) < 9$ Is read: minus seven is **'less than'** nine.

Exercises

1. Write down the number corresponding to each letter:

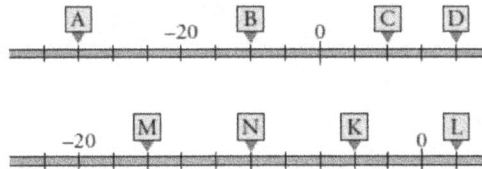

2. Write down the temperatures shown on these thermometers. Find their position in a number line.

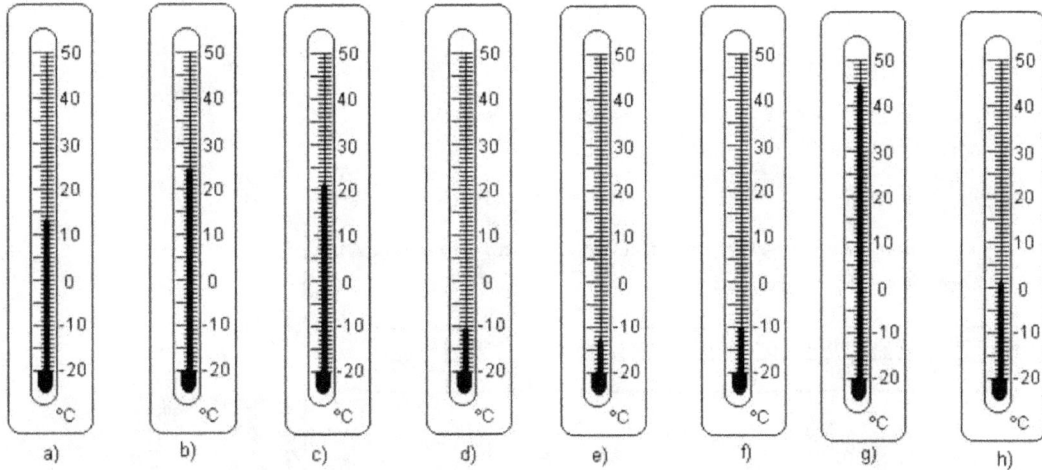

3. What is the temperature which is:

a) 7 degrees lower than 5° C

b) 6 degrees lower than 4° C

c) 16 degrees higher than - 4° C

d) 9 degrees lower than - 6° C

4. What is the difference in temperature between each pair of thermometers?

5. Write using integers.

a) Four degrees above zero

b) A deposit of $200.00

c) 250 meters below sea level

d) Three degrees below zero

e) An elevation of 8848 meters above sea level.

6. Use a number line to correctly place the sign < (greater than) or < (less than) between the numbers in each of the following pairs.

a) 89 ___ 98   b) 89___(-98)   c) 98___(-69)

7. This table shows estimations of the mean temperatures on the surface of nine planets. List the planets on one line in order from hottest to coldest.

Earth	Jupiter	Mars	Mercury	Neptune	Pluto	Saturn	Uranus	Venus
8˚C	−150˚C	−37˚C	179˚C	−225˚C	−236˚C	−185˚C	−214˚C	453˚C

8. Which temperature is lower (colder), 8°C or (-4)°C ?

9. Rearrange the following numbers in order of size, largest first: (-2), 2, 0.5, (-1.5), -8

2. Opposite and Absolute value of a number

The *opposite* of a number is the number on the number line that is the same distance from zero but on the opposite side. To find the opposite of a number, we only have to change its sign.

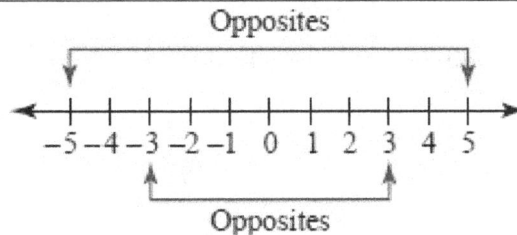

Example 1:

   a) The opposite of 26 is (-26)

   b) The opposite of (-19) is 19.

The **absolute value** of a signed number is the number of units between the number and zero on the number line. Absolute value is defined as the number of units only; direction is not involved. Therefore, the absolute value is never negative.

- If the number is *positive*, the absolute value is the *same* number.

- If the number is *negative*, the absolute value is the *opposite*.

The absolute value of a number is always a positive number (or zero). We specify the absolute value of a number n by writing n in between two vertical bars: $|n|$.

Example 2: Calculate the absolute value of 15, (− 8), 6, (− 10), 0, 123 and (− 3404):

Solution:

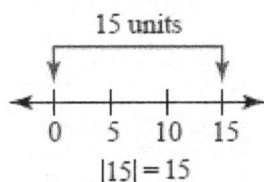
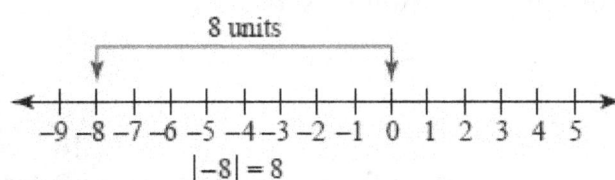

$|6| = 6$ $|(-10)| = 10$ $|0| = 0$ $|123| = 123$ $|(-3404)| = 3404$

10. a) The opposite of __ is 61.   b) The opposite of (-23) is __ .

11. Find the opposite of these numbers:

a) (-42)   b) (-57)   c) 3.78   d) (-0.55)   e) 0.732

12. Find the absolute value of the signed number.

a) $|(-0.25)|$ b) $|-(-6)|$ c) $|\frac{5}{7}|$ d) $|-\frac{2}{3}|$ e) $|0.1657|$

13. A cyclist travels up a mountain 415 m, then turns around and travels down the mountain 143 m. Represent each trip as a signed number.

14. If 80 km north is represented by 80, how would you represent 80 km south?

15. Simplify:

a) $|16 - 10| - |14 - 9|$ b) $8 - |12 - 8| - |10 - 8| + 2$

---

3. Addition of integers

There is a way to understand how to add integers. In order to add positive and negative integers, we will imagine that we are moving along a number line.

If we want to add (-1) and 5, we start by finding the number -1 on the number line, exactly one unit to the left of zero. Then we would move five units to the right. Since we landed four units to the right of zero, the answer must be 4.

If asked to add 3 and (-5), we can start by finding the number three on the number line (to the right of zero).

Then we move five units left from there because negative numbers make us move to the left side of the number line. Since our last position is two units to the left of zero, the answer is (-2).

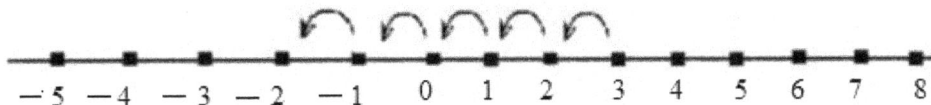

To add two integers,

1. If the signs are alike, add their absolute values and use the common sign for the sum.

2. If the signs are not alike, subtract the smaller absolute value from the larger absolute value. The sum will have the sign of the number with the larger absolute value.

Example 3.

1. If the signs are alike, add their absolute values and use the common sign for the sum.

    a) $5 + 2 = 7$ b) $-5 - 3 = (-8)$ c) $7 + 4 = 11$ d) $-20 - 10 = (-30)$

2. If the signs are not alike, subtract the smaller absolute value from the larger absolute value. The sum will have the sign of the number with the larger absolute value.

    a) $3 - 5 = (-2)$ b) $(-7) + 5 = (-2)$ c) $10 - 8 = 2$ d) $5 - 6 = (-1)$

Exercises

16. Evaluate:

a) $9 - 4$ b) $4 - 9$ c) $10 - 8$ d) $8 - 9$ e) $11 - 7$

f) $7 - 11$ g) $5 - 11$ h) $3 - 7$ i) $1 - 6$ j) $10 - 12$

k) $11 - 15$ l) $14 - 20$

17. Work out:

a) $-2 + 6$ b) $-4 + 7$ c) $-1 + 9$ d) $-7 + 2$ e) $-8 + 5$

f) $-10 + 8$ g) $-12 + 5$ h) $-15 + 6$ i) $-15 + 14$

STOP **No more than one sign (+) or (-) before a number!!!**

In Maths, it is no allowed to write two signs (+) or (-), together before a number. When this happens, we must change both signs by only one. We will do it as follows:

You must pay attention to the first sign:

- If the first sign is *(+)*, then, quit it and *leave* the second one.
- If the first sign is *(-)*, then, quit it and *change* the second by its opposite.

Example 4:

a) $+(-7) = (-7)$ b) $-(-2) = +2$ c) $-(+8) = (-8)$ d) $-(+1) = (-1)$

e) $+(+11) = +11$ f) $+(-14) = (-14)$

18. Add:

a) $1 + 3$ b) $(-8) + 5$ c) $2 + 1$ d) $(-4) + 4$ e) $5 + (-6)$

f) $(-6) + 4$ g) $7 + (-2)$ h) $(-3) + 5$ i) $(-3) + (-2)$ j) $(-4) + (-2)$

k) $(-2) + (-2)$ l) $(-5) + 6$ m) $(-2) + (-5)$ n) $(-5) + 10$

ñ) $(-7) + (-4)$ o) $(-4) + 7$ p) $(-5) + (-2)$ q) $(-7) + (-10)$

19. Complete:

a) $(-11) + \boxed{....} = 4$ b) $(+13) + \boxed{....} = 12$ c) $\boxed{....} + (-20) = (-12)$

d) $(+3) + \boxed{....} = (-7)$ e) $(-15) + \boxed{....} = 9$ f) $\boxed{....} + 8 = 7$

20. Calculate, as shown in the first, made for you:

a) $-(-3) + (-7) = +3 - 7 = (-4)$ b) $3 - (-2)$ c) $15 + (-21)$ d) $(-52) - 24$

e) $2 + (-3)$ f) $-(-3) + (-4)$ g) $(-1) - (-6)$

21. Mount Everest is 29,028 feet above sea level. The Dead Sea is 1,312 feet below sea level. What is the difference of altitude between these two points?

22. The temperature in Chicago was 4° C at two in the afternoon. If the temperature dropped 12° C at midnight, what is the temperature now?

23. A submarine was situated 2100 feet below sea level. If it ascends 1230 feet, what is its new position?

24. Aristotle was born in 384 B.C. and died in 322 BC. How old was he he died?

25. A submarine was situated 1230 feet below sea level. If it descends 125 feet, what is its new position?

26. On the 2nd of January, the temperature dropped from 3°C at two o'clock in the afternoon to -11°C at 8 a. m. the next day. How many degrees did the temperature fall?

99

5. Addition and subtraction of more than two integers

In real life, in Nature, addition and subtractions are related to more than two numbers. In this part of the chapter, you will learn to calculate them.

Imagine you are in the beginning of a very long street and ask someone where your favorite cloths shop is. Imagine now his answer:

"Go by bus to the next stop, which is exactly at 500 meters, come back 100 meters, and come back for 50 meters more. There, you will find your shop".

You can write your walk as follows: $+500 - 100 - 50$

Don´t you find it a bit difficult? It would have been easier: *"Go by bus to the next stop, which is exactly at 500 meters and come back 150 meters".*

When you have to calculate additions and subtractions implying more than two numbers, you must add separately positive numbers, and later, negative numbers. Finally calculate the difference

TOTAL POSITIVES – TOTAL NEGATIVES

To calculate additions and subtractions implying more than two numbers:

1. Add separately positive numbers,

2. Add separately negative numbers,

3. Finally, calculate the difference

TOTAL POSITIVES – TOTAL NEGATIVES

Example 5:   Calculate:

a) $7 - 2 + 4 + 1 - 8$ 　　　　b) $10 + 2 - 5 - 9 + 2$ 　　　　c) $-2 + 5 - 4 - 10 + 1$

Solution:

a) $5 - 2 + 4 + 1 - 8$

    1st: add separately positive numbers $\rightarrow +7 + 4 + 1 = 12$

    2nd: add separately negative numbers $\rightarrow -2 - 8 = -10$

    3rd: finally, calculate the difference $\rightarrow 12 - 10 = \boxed{2}$

b) $10 + 2 - 5 - 9 + 2$

    1st: add separately positive numbers $\rightarrow +10 + 2 + 2 = 14$

    2nd: add separately negative numbers $\rightarrow -5 - 9 = -14$

    3rd: finally, calculate the difference $\rightarrow 14 - 14 = \boxed{0}$

c) $-2 + 5 - 4 - 10 + 1$

    1st: add separately positive numbers $\rightarrow +5 + 1 = 6$

    2nd: add separately negative numbers $\rightarrow -2 - 4 - 10 = -16$

    3rd: finally, calculate the difference $\rightarrow 6 - 16 = \boxed{(-10)}$

---

27. Calculate:

    a) $+2 - 7 + 5$ b) $+12 - 5 - 8$ c) $13 - 9 + 5 - 7$

    d) $6 - 8 - 6 + 5 + 4 - 6$ e) $-3 - 5 + 2 - 1 - 7 + 4$ f) $-8 - 7 + 2 + 9 - 10 + 18$

28. Quit parenthesis and Work out:

    a) $(+3) - (+8)$ b) $(-9) + (-6)$ c) $(-7) - (-7) - (+7)$

    d) $(-11) + (+8) - (-6)$ e) $(+15) - (-12) - (+11) + (-16)$

    f) $(-3) - (-2) - (+4) + (-7) + (+8)$ g) $(+11) - (+7) + (-13) - (-20) + (-11)$

29. Quit parenthesis and Work out:

    a) $-(-3) + (-7) - (-2) + (-9)$ b) $3 - 7 - (-2) + 9$ c) $15 + (-21) - 20 - (-1$

    d) $(-52) - 24$ e) $(-72) + 80 - 8$ f) $2 + (-3) + (-5) + (-4) + 6$

    g) $(-3) + 6 + (-4) + 4 + (-6)$ h) $(-1) - 6 + (-7) - (-5) - 21 + (-12)$

Exercises

6. Addition and subtraction with parentheses

In operations implying parentheses, pay attention:

FIRST OPERATE INTO THE PARENTHESIS!!!

Example 6: Calculate:

a) $13 - (6 + 5)$ b) $(4 + 8) - (3 - 9)$ c) $10 + (8 - 15 + 2 - 6)$

Solutions:

a) $13 - (6 + 5)$

First, operate into the parentheses: $(6 + 5) = 11$

Then, substitute in the main operation line: $13 - 11 = \boxed{2}$.

b) $(4 + 8) - (3 - 9)$

First, operate into the parentheses: $\begin{cases} (4+8) = 12 \\ (3-9) = (-6) \end{cases}$

Then, substitute in the main operation line: $12 - (-6) = 12 + 6 = \boxed{8}$.

c) $10 + (8 - 15 + 2 - 6)$

First, operate into the parentheses: $(8 - 15 + 2 - 6) =$

1st: add separately positive numbers $\rightarrow +8 + 2 = 10$

2nd: add separately negative numbers $\rightarrow -15 - 6 = -21$

3rd: finally, calculate the difference $\rightarrow 10 - 21 = (-11)$

Then, substitute in the main operation line: $10 + (-11) = 10 - 11 = \boxed{(-1)}$.

30. Calculate:

    a) $8 - (6 + 5)$ b) $12 - (7 + 11 - 14 - 8)$ c) $(6 - 12 + 2) - (11 - 4 + 2 - 5)$

31. Calculate:

    a) $(5 - 7) - [(-3) + (-6)]$ b) $(-8) + [(+7) - (-4) + (-5)]$

    c) $(+9) - [(+3) - (3 - 12) - (+8)]$ d) $[(+6) - (-8)] - [(-4) - (-10)]$

    e) $[(2 - 8) + (5 - 7)] - [(-9 + 6) - (-5 + 7)]$

7. Multiplying and dividing integers

When multiplying or dividing integers, we will multiply or divide their absolute values (in the same way as when multiplying or dividing natural numbers). But, after that, we will have to pay attention to their signs, because:

Product of integers of the **same sign** will give a **positive number**.

Product of integers with **opposite sign** will give a **negative number**.

If any of the integers in the product is 0, the product is 0.

* And, the same rules about signs are applied in divisions.

Example 7: Calculate:

    a) 4×3 b) $(-4) \times (-5)$ c) $(-7) \times 6$ d) $12 \times (-2)$

    e) $14 \div 2$ f) $(-24) \div (-3)$ g) $(-100) \div 25$ h) $98 \div (-7)$

Solution:

    a) 4×3 Both numbers are positive (same sign), so their product is a positive number, 12.

    b) $(-4) \times (-5)$ Both numbers are negative (same sign), so their product is a positive number, 20.

    c) $(-7) \times 6$, the first number is negative and the second is positive, so the product is a negative number, (-42).

d) 12×(-2) the first number is positive and the second is negative, so the product is a negative number, (-24).

e) 14 ÷ 2 Both numbers are positive, so the quotient is a positive number, 7.

f) (-24) ÷ (-3) Both numbers are negative, so the quotient is a positive number, 8.

g) (-100) ÷ 25 The numbers have different signs, so the quotient is a negative number, (- 4).

h) 98 ÷ (-7) Both numbers have different signs, so the quotient is a negative number, (-14).

32. Multiply:

a) $5 \cdot 4$ b) $(-3) \cdot (-8)$ c) $3 \cdot (-5)$ d) $(-4) \cdot 5$ e) $(-7) \cdot (-6)$

f) $5 \cdot 7$ g) $(-4) \cdot 8$ h) $9 \cdot (-5)$ i) $(-11) \cdot 7$ j) $10 \cdot (-10)$

k) $(-8) \cdot 5$ l) $(-10) \cdot (-10)$

33. Multiply:

a) $(-3) \cdot (-5)$ b) $(-4) \cdot (-2)$ c) $9 \cdot (-6)$ d) $6 \cdot (-3)$ e) $9 \cdot (-9)$

f) $5 \cdot (-8)$ g) $(-5) \cdot (-8)$ h) $(-7) \cdot 9$ i) $9 \cdot (-5)$ j) $(-4) \cdot (-7)$

k) $(-5) \cdot 5$ l) $(-10) \cdot (-2)$

34. Multiply:

a) $(+21) \cdot (+3) \cdot (+4)$ b) $(+19) \cdot (-2) \cdot (+3)$

c) $(+13) \cdot (-5) \cdot (-6)$ d) $(-20) \cdot (-9) \cdot (-3)$

35. Multiply:

a) $(-5) \cdot \square = (-30)$ b) $\square \cdot (+3) = 45$ c) $(-9) \cdot \square = 27$ d) $\square \cdot (-8) = (-48)$

36. Divide:

a) $(-15) : (-5)$ b) $(-4) : (-2)$ c) $9 : (-3)$ d) $6 : (-3)$

e) $9 : (-9)$ f) $16 : (-8)$ g) $(-40) : (-8)$ h) $(-63) : 9$

37. Divide:

    a) $25 : (-5)$ b) $(-21) : (-7)$ c) $8 : (-4)$ d) $4 : (-1)$

    e) $8 : (-4)$ f) $(-24) : (-3)$ g) $(+12) : 2$ h) $72 : (-8)$

38. Copy and complete:

    a) $\square : 4 = -10$ b) $40 : \square = (-8)$ c) $(-100) : \square = -25$

    d) $\square : (-12) = 6$

39. Copy and complete:

    a) $(-36) : \square = (-4)$ b) $(-54) : \square = 9$ c) $\square : (-6) = (-42)$

    d) $48 : \square = (-6)$ e) $(-63) : \square = (-7)$ f) $\square : 8 = 2$

---

8. Order of operations

To evaluate an expression with more than one operation
Step 1. Parentheses   **Step 2.** Exponents
Step 3. Multiply and Divide   **Step 4.** Add and Subtract

Example 8: Calculate:

    a) $2 \cdot 7 - 3 \cdot 4 - 2 \cdot 3$ b) $30 : 6 - 42 : 7 - 27 : 9$ c) $16 + (-5) \cdot 4$ d) $20 - (-6) \cdot (-4)$

Solution:

a) $2 \cdot 7 - 3 \cdot 4 - 2 \cdot 3$ $\rightarrow$ No parentheses, no exponents, so, we go directly to step 3, multiplications:

$$2 \cdot 7 = 14$$
$$3 \cdot 4 = 12$$
$$2 \cdot 3 = 6$$

And substitute: $14 - 12 - 6 = 14 - 18 = \boxed{(-4)}$.

b) $30 : 6 - 42 : 7 - 27 : 9 \quad \rightarrow \quad$ No parentheses, no exponents, so, we go directly to step 3, divisions:

$$30 : 6 = 5$$

$$42 : 7 = 6$$

$$27 : 9 = 3$$

And substitute: $5 - 6 - 3 = 5 - 9 = \boxed{(- 4)}$.

c) $16 + (-5) \cdot 4 \quad \rightarrow \quad$ No parentheses, no exponents, so, we go directly to step 3, multiplication:

$$(-5) \cdot 4 = (-20)$$

And substitute: $16 + (-20) = \boxed{(- 4)}$.

d) $20 - (- 6) \cdot (-4) \quad \rightarrow \quad$ No parentheses, no exponents, so, we go directly to step 3, multiplication:

$$(- 6) \cdot (- 4) = 24$$

And substitute: $20 - 24 = \boxed{(- 4)}$.

---

40. Calculate. Pay attention to order of operations:

a) $3 \cdot 5 - 4 \cdot 6 + 5 \cdot 4 - 6 \cdot 5$

b) $5 \cdot 4 - 28 : 4 - 3 \cdot 3$

c) $(-2) \cdot (-5) + (+4) \cdot (-3)$

d) $(-8) \cdot (+2) - (+5) \cdot (-4)$

e) $10 + (- 4) \cdot (+2) - (+6)$

f) $(-5) - (+4) \cdot (-3) - (-8)$

g) $14 - (+5) \cdot (- 4) + (- 6) \cdot (+3) + (-8)$

h) $(+4) \cdot (- 6) - (-15) - (+2) \cdot (-7) - (+12)$

Exercises

Example 9: Calculate:

    a) $60 : (8 - 14) + 12$ b) $(9 - 13 - 6 + 9) \cdot (5 - 11 + 7 - 4)$

Solutions:

    a) $60 : (8 - 14) + 12$

        Step 1. Parentheses $\rightarrow$ $8 - 14 = (-6)$

        Step 2. Exponents $\rightarrow$ There are not. $\rightarrow$ So we have: $60 : (-6) + 12$

        Step 3. Division $\rightarrow$ $60 : (-6) = (-10)$

        Step 4. Add and Subtract $\rightarrow$ $(-10) + 12 = \boxed{(-2)}$.

    b) $(9 - 13 - 6 + 9) \cdot (5 - 11 + 7 - 4)$

        Step 1. Parentheses $\rightarrow$ $\begin{cases} (9 - 13 - 6 + 9) = 18 - 19 = (-1) \\ (5 - 11 + 7 - 4) = 12 - 15 = (-3) \end{cases}$

        Step 2. Exponents $\rightarrow$ There are not. $\rightarrow$ So we have: $(-1) \cdot (-3)$

        Step 3. Product $\rightarrow$ $(-1) \cdot (-3) = \boxed{3}$.

Exercises

41. Calculate. Pay attention to order of operations:

    a) $3 \cdot (3 - 5)$ b) $4 \cdot (8 - 6)$ c) $5 \cdot (8 - 12)$

    d) $(-2) \cdot (7 - 3)$ e) $(-4) \cdot (6 - 10)$ f) $(-5) \cdot (2 - 9)$

    g) $16 : (1 - 5)$ h) $(-35) : (9 - 2)$ i) $(-14) : (5 + 2)$

    j) $(2 - 8) : 3$ k) $(5 + 7) : (-4)$ l) $(12 - 4) : (-2)$

42. Calculate. Pay attention to order of operations:

    a) $35 + 7 \cdot (6 - 11)$ b) $(6 + 2 - 9 - 15) : (7 - 12 + 3 - 6)$

    c) $-(8 + 3 - 10) \cdot [(5 - 7) : (13 - 15)]$

43. Calculate. Pay attention to order of operations:

a) $(-3) \cdot [(-9) - (-7)]$

b) $28 : [(-4) + (-3)]$

c) $[(-9) - (+6)] : (-5)$

d) $(-11) - (-2) \cdot [15 - (+11)]$

e) $(+5) - (-18) : [(+9) - (+15)]$

f) $(-4) \cdot [(-6) - (-8)] - (+3) \cdot [(-11) + (+7)]$

Review exercises

1. Rearrange the following numbers in order of size, largest first: $(-2), 2, 0.5, (-1.5), -8$

a) $6, 2, 0, 4, (-7), 3$

b) $(-7), -(2,) 0, (-1), (-5), (-9)$

c) $(-4), 0, 6, (-8), 3, (-5)$

2. Evaluate:

a) $-1 - 1$

b) $-1 - 2$

c) $-2 - 3$

d) $-2 - 5$

e) $-4 - 3$

f) $-7 - 1$

g) $-6 - 6$

h) $-10 - 2$

i) $-3 - 12$

3. Calculate the following sums of integers:

a) $5 + (-9)$

b) $5 + (-11)$

c) $13 + (-9)$

d) $22 + (-30)$

e) $21 + (-33)$

f) $46 + (-52)$

g) $(-8) + (-14)$

h) $(-21) + (-15)$

i) $(-33) + (-22)$

j) $(-13) + 18$

k) $(-22) + 9$

l) $(-37) + 21$

4. Calculate:

a) $4 + (-3)$

b) $8 - (-5)$

c) $4 + (-2) - (-3) + 5$

d) $5 - (-2) + 6 - (-4)$

e) $(-5) - (-3) + 10 - (-2)$

f) $4 + (-7) - (-2) - 3$

g) $(-8) + (-3) - (-1) + (-4)$

h) $(-9) - 12 - 1$

5. Calculate:

a) $-15 + 7 + (-8) + 9$

b) $15 + (-10) + 5 + (-10)$

c) $-12 + 8 + (-7) + (-2)$

d) $-13 + 15 + (-10) + 9$

e) $-20 - (-8 + 4 - 5)$

f) $12 - (-8 + 10)$

g) $-(-8) - (4 - 7 - 9) =$

h) $-15 - (-2 + 9)$

6. Multiply or divide:

   a) $4 \cdot (-3)$ b) $(-5) \cdot 2$ c) $3 \cdot (-7)$ d) $(-2) \cdot (-5)$

   e) $3 \cdot (-4)$ f) $-5 \cdot 3$ g) $4 \cdot (-2) \cdot 3$ h) $(-3) \cdot (-2) \cdot 4$

   i) $(-1) \cdot (-4) \cdot (-5)$ j) $(-7) \cdot 2 \cdot (-3)$ k) $(-2) \cdot 4 \cdot (-3)$ l) $(-12) : 4$

7. Multiply or divide:

   a) $24 : (-6)$ b) $(-8) : (-2)$ c) $(-27) : (-1)$ d) $(-12) : (-4)$

   e) $18 : (-3)$ f) $(-6) \cdot (-2) : (-4)$ g) $3 \cdot (-8) : (+6)$ h) $6 \cdot (-8) : (-12)$

   i) $(-7) \cdot 6 : (-21)$ j) $4 \cdot (-9) : 12$

8. Calculate. Pay attention to order of operations:

   a) $8 + (4 - 9 + 7) \cdot 2 + 4 \cdot (3 - 8 + 4)$

   b) $4 \cdot [(+5) + (-7)] - (-3) \cdot [7 - (+3)]$

   c) $(-3) \cdot (+11) - [(-6) + (-8) - (-2)] \cdot (+2)$

   d) $(-6) \cdot [(-7) + (+3) - (7 + 6 - 14)] - (+7) \cdot (+3)$

   e) $[(+5) - (+2)] : [(-8) + (-3) - (-10)]$

Unit 7.- Proportionality and percentages

1. Introduction to proportionality

From the earliest times, humans have drawn maps to represent the geography of their surroundings. Some maps depict features encountered on a journey, like rivers and mountains.

The most useful maps incorporate the concept of scale, or proportion. Simply put, a scaled map accurately preserves relative distances. So if the distance from one city to another is twice the distance from the city to a river in real life, the distance between the cities is twice the distance from the city to a river on the map as well.

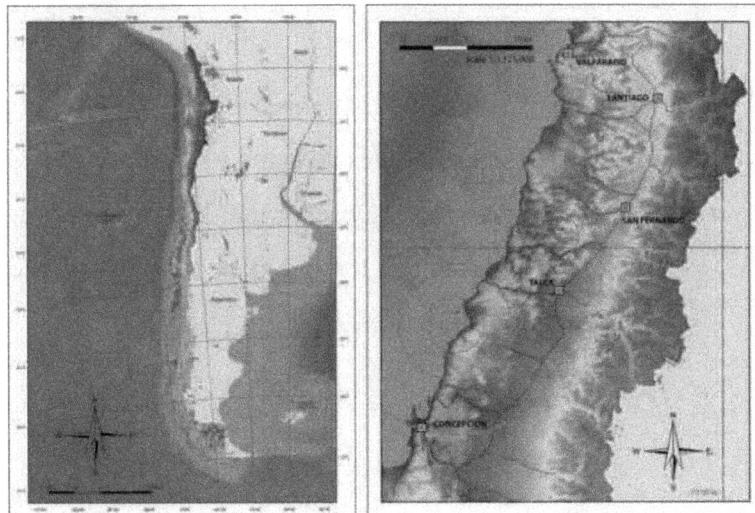

2. Direct and inverse proportionality

We say that there is a **direct proportionality** between two magnitudes if when we double one magnitude, the other also doubles, when we half the first, the second also halves.

We say that there is an **inverse proportionality** between two magnitudes if when increasing one magnitude, (double, triple…) the other decreases (half, third…), when decreasing one (half, third…), the other increases (double, triple…).

Example 1: Direct and inverse proportionality examples.

a) If a can of cola costs 40 cent, the cost of:
- 2 cans is 80 cent
- 5 cans is € 2.00 (200 cent).

We can see that, for example, if we double the number of cans, we double the price. They are in **direct proportion**.

b) Look at the relationship that exists between the number of the members of a family and the days that one box of apples lasts them (suppose that all the people eat the same amount of apples at the same rate). Observe that the more people there are in the family the less time the box of fruit lasts, and the less people there are, the longer it lasts. These two magnitudes are in **inverse proportion**.

1. Classify the relationship of proportionality (direct, inverse or non-proportional) between the following pairs of magnitudes:

   a) The number of children attending a birthday party and the size of the piece of a cake that corresponds to each.

   b) The number of workers repairing a street and the number of days it takes them.

   c) Number of people attending a doctor for an hour and the time spent with each patient.

   d) The width of the shelf and the number of books (same type) you can put.

   e) Hours of operation of a machine and number of parts produced.

   f) The ability of a deposit and the time we need to fill it, using the same supplier.

   g) Megabytes capacity of a pen-drive and the number of photos you can store on it.

   h) People who lift an object and the strength they must do to load it.

2. Classify the relationship of proportionality (direct, inverse or non-proportional) between the following pairs of magnitudes:

   a) The number of kilos sold and the money raised.

   b) The number of workers who do work and time spent.

   c) A man's age and his height.

   d) The speed of a vehicle and the distance traveled in half an hour.

   e) The time it stays open a tap and the amount of water that sheds.

   f) The flow of a tap and the time it takes to fill a reservoir.

   g) The number of pages of a book and its price.

3. Direct proportionality problems

How are proportionality problems solved? The method to solve proportionality problems depends on the proportionality relationship between the magnitudes comparing.

First of all, we are going to have to build a table with the data we have. So, we are going to have two fractions and we will have to build an equality with both fractions, *more or less*. We are going to see it with an example.

Example 2: If 3 kg of oranges cost € 2.64, what is the cost of 20 kg of them?

Solution: First of all, we are building a table with the data we have:

kg of oranges:	D	Euros:
3		2.64
20		X

We have also inserted a letter D between the two magnitudes we are comparing (kg and euros), meaning they are in direct proportion.

Now, we have two fractions, $\frac{3}{20}$ *and* $\frac{2.64}{x}$. As they are in **direct proportion**, we are building an equality with them, but writing at the left of the "=" sign, the fraction containing the "x":

$$\frac{2.64}{x} = \frac{3}{20}$$

As these two fractions are equivalent, we already know how to calculate "x", by "Cross Multiplying".

$$\frac{2.64}{x} = \frac{3}{20} \quad \Rightarrow \quad 3 \cdot x = 20 \cdot 2.64$$
$$3x = 52.8$$
$$x = \frac{52.8}{3} = 17.6 \; euros$$

Exercises

3. If 3 liters of petrol cost 3.45 €. How much will cost

a) 5 liters?

b) 23.5 liters?

114

4. If we travel 136 km in 1.5 hours driving at a constant speed.

a) How many km will we travel in 7.4 h?

b) How many hours do we need to travel 200 km?

5. Adrian finds that in each delivery of 500 bricks there are 20 broken bricks. How many bricks are broken in a delivery of 7500?

6. In a drink, 53 ml of fruit juice are mixed with 250 ml of water. How many liters of water are there in 30 l of that drink?

7. A car uses 25 liters of petrol to travel 176 miles. How far will the car travel using 44 liters of petrol?

8. Four kilos of potatoes cost € 1.80. How much do three kilos cost?

9. A machine produces 800 screws in 5 hours. How long will it take to produce 1000 screws?

10. A family drinks 2.5 L of milk daily. How many liters are consumed per week?

11. Complete in your notebook: If you need three baskets to take 15 melons, how many baskets will you need to take 195 melons?

12. A NGO delivers every day 48,000 kg of food using its 4 trucks. How many kg can they deliver one day if one of the trucks has broken down?

13. In a car factory, 380 cars have been built in 5 hours. How many cars will be produced in 12 hours, keeping the same ratio?

4. Inverse proportionality problems

The method to solve **inverse** proportionality problems has the same initial steps than direct proportionality method. It only has a little difference, before building the equality with the fractions.

We are going to see it with an example:

Example 3: Eight workers have to repair a street, and it takes them 90 minutes. Tomorrow, they must repair another similar street, but they are going to be 9 workers. How long will it take them?

Solution: First of all, we are building a table with the data we have:

Number of workers:	I	Number of minutes:
8		90
9		X

We have also inserted a letter I between the two magnitudes we are comparing (workers and minutes), meaning they are in inverse proportion.

Now, we have two fractions, $\dfrac{8}{9}$ *and* $\dfrac{90}{x}$. Now, we are building an equality with them, writing at the left of the *"="* sign, the fraction containing the *"x"*:   As they are in **inverse proportion**, we will *"UPSIDE-DOWN"* the other fraction (at the right of the *"="* sign).

$$\frac{90}{x}=\frac{9}{8}$$

As these two fractions are equivalent, we already know how to calculate *"x"*, by *"Cross Multiplying"*.

$$\frac{90}{x}=\frac{9}{8} \quad \Rightarrow \quad 9\cdot x = 90\cdot 8$$
$$9x = 720$$
$$x = \frac{720}{9} = 80 \ \text{min}$$

15. A truck that carries 3 tons need 15 trips to carry a certain amount of sand. How many trips are needed to carry the same amount of sand with another truck that carries 5 tons?

16. An automobile factory produces 8100 vehicles in 60 days. With the production rhythm unchanged. How many units will be made in one year?

17. A driver takes 3.5 hours to drive 329 km. How long will it take another trip in similar conditions as the previous one, but travelling 282 km instead?

18. Two hydraulic shovels make the trench for a telephone cable in ten days. How long will it take to make the trench with 5 shovels?

19. A farmer has grass to feed 20 cows for 60 days. If he had 30 cows, for how many days could he feed them?

20. At a fountain, it took 24 seconds to fill a 30-liter jar. How long does it take to fill a 50 liter container?

21. A pool has three similar taps. If opened two of them, pool gets full in 45 minutes. How long will it take filling it if you open all three?

22. Ten gardeners took eight days to repair all the trees in the streets of a town. How much would it have taken if they had been four gardeners only?

23. A gardener, working 8 hours per day, prepares his field in nine days. How long will it take him doing the same work, if he worked 12 hours a day?

24. Eight people collect all the oranges from an orchard in 9 hours. How long would it take them if they were 6 people?

We have already studied the proportionality relationship between two magnitudes. In some situations, there are three or more magnitudes having a proportionality relationship. These cases are named **composite proportionality**. Direct and inverse proportionality are both named *simple proportionality*.

How are composite proportionality problems solved?

The method to solve **composite** proportionality problems has lots of steps in common with simple proportionality one.

We are going to see it with an example.

Example 4: 50 cows consumed 4200 kg of grass in 7 days. How many kg of grass are needed to feed 20 cows for 15 days?

Solution: First of all, we are building a table with the data we have:

Cows:	Kg of grass:	Days:
50	4200	7
20	X	15

Now, we have to indicate the proportionality relationship of the magnitude containing the "*x*", with each one of the others. We use two lines starting at the magnitude with the "*x*" and going to the others, writing on each line, the kind of proportionality between them:

Cows:	Kg of grass:	Days:
50	4200	7
20	X	15

Now, we have three fractions, $\dfrac{50}{20}$, $\dfrac{4200}{x}$ and $\dfrac{7}{15}$. We are building an equality with them.

In composite proportionality problems, we write at the left of the "=" sign, the fraction containing the "x". Later, at the right of the "=" sign, we will write the product of the other two fractions, but writing them as in simple proportionality problems:

- If it is Direct proportional, we write the fraction.
- If it is Inverse proportional, we write *"UPSIDE-DOWN"* the fraction.

In our case:

$$\frac{4200}{x} = \frac{50}{20} \cdot \frac{15}{7}$$

Now, we can multiply the fractions at the right of the "=" sign. SIMPLIFY IF POSSIBLE!!!

$$\frac{50}{20} \cdot \frac{15}{7} = \frac{5}{2} \cdot \frac{15}{7} = \frac{\cdot 5 \cdot 3 \cdot 5}{2 \cdot 7} = \frac{75}{14}$$

And now: $\quad \dfrac{4200}{x} = \dfrac{75}{14}$

As these two fractions are equivalent, we already know how to calculate "x", by *"Cross Multiplying"*.

$$\frac{4200}{x} = \frac{75}{14} \quad \Rightarrow \quad 75 \cdot x = 4200 \cdot 14$$

$$75x = 58800$$

$$x = \frac{58800}{75} = 784 \ kg$$

Example 5: Five gardeners, working 12 hours a day, complete their gardening work in 10 days. How many gardeners would have been necessary to finish the same work, working 10 hours each day during 6 days?

<u>Solution</u>: First of all, we are building a table with the data we have:

Gardeners:	Hours per day:	Days:
5	12	10
X	10	6

Now, we have to indicate the proportionality relationship of the magnitude containing the *"x"*, with each one of the others. We use two lines starting at the magnitude with the *"x"* and going to the others, writing on each line, the kind of proportionality between them:

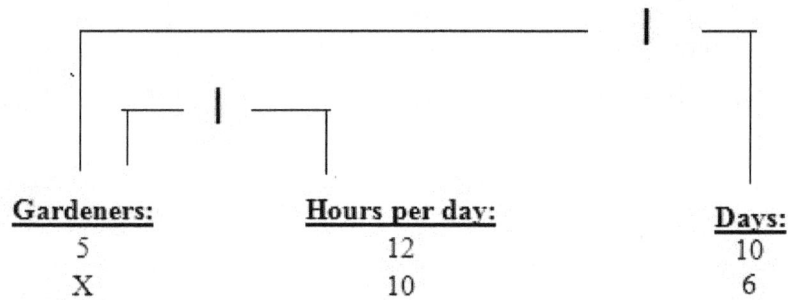

Gardeners:	Hours per day:	Days:
5	12	10
X	10	6

Now, we have three fractions, $\dfrac{5}{x}, \dfrac{12}{10}$ *and* $\dfrac{10}{6}$. We are building an equality with them.

In our case, both of the relationships are Inverse proportional, so, we write *"UPSIDE-DOWN"* both of them:

$$\frac{5}{x} = \frac{10}{12} \cdot \frac{6}{10}$$

Now, we can multiply the fractions at the right of the *"="* sign.   SIMPLIFY IF POSSIBLE!!!

$$\frac{10}{12} \cdot \frac{6}{10} = \frac{6}{12} = \frac{2 \cdot 3}{2 \cdot 2 \cdot 3} = \frac{1}{2}$$

And now: $\dfrac{5}{x} = \dfrac{1}{2}$

As these two fractions are equivalent, we already know how to calculate *"x"*, by *"Cross Multiplying"*.

$$\frac{5}{x} = \frac{1}{2} \quad \Rightarrow \quad 1 \cdot x = 5 \cdot 2$$
$$x = 10 \, people$$

25. In a manufacturing workshop, with 6 sewing machines, 600 jackets were manufactured in 10 days. How many jackets would be manufactured with 5 machines in 15 days?

26. An industrial washer, working 8 hours a day for five days, has washed 1000 kg of clothes. How many kg of clothes will be washed in 12 days working 10 hours a day?

27. Five interviewers, working 8 hours a day, complete data for a market survey in 27 days. How long will it take them doing the same work if they were 9 people working 10 hours each day?

28. The garden of a park has been cleaned by 3 gardeners working for 15 days and working 8 hours daily. In another park, they want to create another similar garden having the same extension. As they want to finish in 4 days, they have called nine gardeners. How many hours must work daily?

6. Percentages

A *percentage (or percentage)* is a ratio of a number to 100. Percentage is expressed using the symbol %.

A percentage is also equivalent to a fraction with denominator 100.

For example, 65% is equivalent to fraction $\frac{65}{100}$.

6.1. Percentage comparison of two numbers

To find the percentage comparison of two numbers

1. Write the ratio of the first number to the base number.
2. Multiply by 100 this fraction and add the symbol %.

Example 6: What percentage of the region is shaded?

a)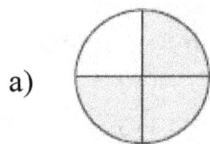

b)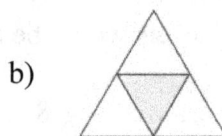

Solution: a) $\dfrac{3}{4}\cdot 100 = \dfrac{3\cdot 100}{4} = \dfrac{300}{4} = \boxed{75\%}$.

b) $\dfrac{1}{4}\cdot 100 = \dfrac{1\cdot 100}{4} = \dfrac{100}{4} = \boxed{25\%}$.

29. What percentage of the region is shaded?

a)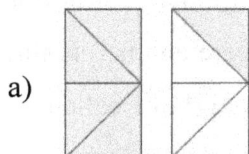

b)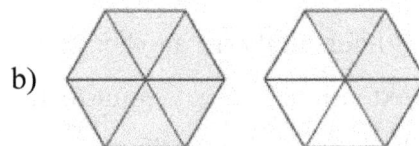

30. Write an exact percentage for these comparisons. Some of them might be calculated directly:

a) 62 of 100 b) 52 per 100 c) 32 to 100 d) 37 to 100 e) 28 per 50

f) 17 per 50 g) 12 of 25 h) 21 to 25 i) 11 per 20 j) 13 per 20

k) 13 to 10 l) 450 to 120 m) 313 of 313 n) 92 to 92 ñ) 30 to 12

o) 44 to 16 p) 85 to 200 q) 65 to 200 r) 15 per 40 s) 83 per 500

t) 70 per 80 u) 98 per 80.

31. It is estimated that 2% of the U.S. population has red hair. This indicates that ___ out of 100 people are redheads.

32. In a recent election there was a 73% turnout of registered voters. This indicates that ___ out of 100 registered voters turned out to vote.

33. In a recent mail-in election, 18 out of every 100 eligible voters did not participate. What percentage of the eligible voters exercised their right to vote?

34. Of the people who use mouthwash daily, 63 out of 100 report fewer cavities. Of every 100 people who report, what percentage do not report fewer cavities?

35. At a football game, 22 children are among the first 100 fans to enter. What percentage of the first 100 fans are children?

36. At the last soccer match of the season, of the first 100 tickets sold, 77 were student tickets. What percentage were student tickets?

37. Write the ratio of 8 to 5 as a percentage.

38. During a campaign to lose weight, the 180 participants lost a total of 4158 lb. If they weighed collectively 37,800 lb before the campaign, what percentage of their weight was lost?

39. James has $500 in his savings account. Of that amount, $35 is interest that was paid to him. What percentage of the total amount is the interest?

40. According to the U. S. Census Bureau, in 2007 one out of every three women aged 25 to 29 had a bachelor's degree or higher. What percentage of women 25 to 29 had a bachelor's degree?

41. According to the U. S. Census Bureau, in 2007, 13 out of every 50 men aged 25 to 29 had a bachelor's degree or higher. What percentage of men 25 to 29 had a bachelor's degree?

42. Mickie bought a TV and makes monthly payments on it. Last year, she paid a total of $900. Of the total that she paid, $180 was interest. What percentage of the total was interest?

7. Solving problems with percentages

To solve problems involving percentages, we will use the following expression:

$$Portion = \frac{Percent}{100} \cdot Whole$$

Anyway, problems with percentages can also be solved as a direct proportionality. We are seeing it in the following examples.

7.1. Calculating the portion

Example 7: What is the 35% of 300?

Solution: We will solve it by applying the percentage expression and as a direct proportionality.

Method 1: Percentage expression:	**Method 2:** As a direct proportionality:
- Portion = x? Calculate: - Percentage = 35 - Whole = 300 $x = \dfrac{35 \cdot 300}{100} = \boxed{105}$. $x = \dfrac{35}{100} \cdot 300$	**Whole:** D **Portion:** 100 35 300 x $\dfrac{100}{300} = \dfrac{35}{x} \rightarrow x \cdot 100 = 300 \cdot 35; x = \dfrac{35 \cdot 300}{100} = \boxed{105}$.

Exercises

43. Calculate:

a) 60% of 80   b) 30% of 91   c) 55% of 72   d) 130% of 90

e) 140% of 70   f) 17.5% of 70   g) 57.5% of 110   h) 45.5% of 80

i) 17.2% of 55   j) 6.14% of 350   k) 12.85% of 980   l) 16 % of 3522

7.2. Calculating the whole

Example 8: 54% of what number is 108?

Solution: We will solve it by applying the percentage expression and as a direct proportionality.

Method 1: Percentage expression:	**Method 2:** As a direct proportionality:
- Portion = 108 Isolate x: - Percentage = 54 - Whole = x? $108 \cdot \dfrac{100}{54} = x = \dfrac{10800}{54} =$ $108 = \dfrac{54}{100} \cdot x$ $= \boxed{200}$.	**Whole:** D **Portion:** 100 54 x 108 $\dfrac{100}{x} = \dfrac{504}{108} = \dfrac{1}{2} \rightarrow x \cdot 1 = 2 \cdot 100 = \boxed{200}$.

44. Calculate and complete:

a) 80% of _____ is 32.   b) 39% of _____ is 39.   c) 497.8 is 76% of ____ .

d) 162 is 18% of ____.   e) 39% of ____ is 105.3.   f) 73% of ____ is 83.22.

g) 124% of ____ is 328.6.   h) 205% of ____ is 750.3.

All together. Calculating percentages, portions and wholes

45. What percentage of 85 is 41? Round to the nearest tenth of a percentage.

46. What percentage of 666 is 247? Round to the nearest tenth of a percentage

47. Eighty-two is 24.8% of what number? Round to the nearest hundredth.

48. Forty-one is 35.2% of what number? Round to the nearest hundredth.

49. Thirty-two and seven tenths percent of 695 is what number?

50. Seventy-three and twelve hundredths percent of 35 is what number?

51. Thirty-seven is what percentage of 156? Round to the nearest tenth of a percentage.

52. Two hundred thirty-two is what percentage of 124? Round to the nearest tenth of a percentage.

53. The cost of a certain model of Ford is 120% of what it was 5 years ago. If the cost of the automobile 5 years ago was $20,400, what is the cost today?

54. In a 1st ESO group, 10 students played football, 15 basketball and 5 tennis. What is the percentage of the students playing football?

55. In a study of 615 people, 185 said they jog for exercise. What percentage of those surveyed jog? Round to the nearest whole percentage.

56. Based on a survey, approximately 77% of TV sets in the United States receive cable. If there are about 304,000,000 sets in the United States, how many do not get cable? Round to the nearest million.

57. The town of Verboort has a population of 17,850, of which 48% is male. Of the men, 32% are 40 years or older. How many men are there in Verboort who are younger than 40?

Exercises

7.3. Percentage changes

As you have seen, percentage changes can be calculated by using previous expression. Anyway, there is an easier expression you can use. It is the following:

Expression to calculate percentage changes:

$$F.A. = I.A. \cdot \left(1 \pm \frac{\%}{100}\right)$$

Where: - F.A.: Final amount.

- I.A.: Initial amount.

- %: Percentage change.

- ±: Use (+) when you consider an increase and a (-) in decreases.

Example 9: Prices of houses have increased 20% from five years ago. In that moment, I paid 120,000 euros for my house. How much would I have to pay today?

Solution:

$\% = 20$;   I.A. $= 120,000$;   F.A. $= x$?;   plus or minus? As prices have increased, we use <u>plus</u> $\rightarrow$

$$F.A. = 120000 \cdot \left(1 + \frac{20}{100}\right) = 120000 \cdot (1 + 0.2) = 120000 \cdot 1.2 = \boxed{144,000 \text{ euros}}.$$

Example 10: In the shop where I am used to buying my clothes, there are 15% sales today. I have paid 25.5 euros for T-shirt. What was its price before sales?

Solution:

$\% = 15$; F.A. $= 25.5$;   I.A. $= x$?; plus or minus? As prices have decreased, we use <u>minus</u> $\rightarrow$

$$F.A. = I.A. \cdot \left(1 \pm \frac{\%}{100}\right) \rightarrow 25.5 = I.A. \cdot \left(1 - \frac{15}{100}\right); \quad 25.5 = I.A. \cdot 0.85; \quad \frac{25.5}{0.85} = I.A. = \boxed{30 \text{ } euros}$$

Example 11: When John began to work in his company, his salary was 1100 euros per month. From that moment, his salary has increased twice, first, 15%, and later, 20%. But, this year, as there are problems with Spanish economy, he has had his salary decreased 35%. What is his salary today?

Solution: We might think *"it increased 15 + 20 = 35%, and it decreased 35%, so, he should have the same initial salary"*.

We are going to see this is not true. As there are more than one percentage change, we will use more than one parentheses in above expression:

$$F.A. = 1100 \cdot \left(1 + \frac{20}{100}\right) \cdot \left(1 + \frac{15}{100}\right) \cdot \left(1\frac{35}{100}\right) = 1100 \cdot 1.2 \cdot 1.15 \cdot 0.65 = \boxed{986.7 \text{ euros}}.$$

You can see his salary has decreased. Be careful in the future, don't forget this!!!

Exercises

58. I am taking part in a program to lose weight. I have already lost a 20%. If my weight now is 60 kg, what was my weight before the program?

59. Today there are 20% sales in a shop near my house. This was a T-shirt's label: "Before 25 €; Today 22.5 €". Do you think this label is correct? Why?

60. During ten last years, prices of houses, first increased 10%, later increased 20% and, this year, they have decreased 30%. Ten years ago, I paid 150,000 € for my house. Check if its price has changed from them. Has it increased or decreased?

Review exercises

1. Classify the relationship of proportionality (direct, inverse or non-proportional) between the following pairs of magnitudes:

    a) Speed and time at a constant speed motion.

    b) Space and time in motion with constant velocity.

    c) Number of persons sharing a cake and size of portion corresponding to each one.

    d) Number of hours a student watches television and hours of study.

    e) Amount of money that you save and your weight.

    f) Number of correct answers and number of failures in an exam.

    g) Number of workers and time it takes to build a wall.

    h) Number of people and amount of food eaten.

    i) Number of people involved in buying a gift and money they bring.

    j) Number of laborers and time taken to finish their work.

2. Complete the table of the cost in € and the liters of petrol bought.

Petrol in litres	1		5	7	
Cost in €		40	6,2		50

3. A phone call costs € 0.25 each 2 minutes or fraction rounded to the seconds. Complete the table.

Call length in minutes	7,5		13 minutes 25 seconds
Cost in €		12	

4. Richard earns £17.5 for working 7 hours. How much will he earn for working 9 hours?

5. Two pumps take 5 days to empty a pool. How long will 5 pumps take to empty the same pool?

6. We have paid for 7 nights in "Hotel Los Llanos" 364 €. How much will we pay for 3 nights? How much for 15 nights?

128

7. It takes 12 hours for 3 bricklayers to build a wall. How long will it take for 5 bricklayers?

8. A 25 kg tin of paint covers 70 m^2 of wall. How many kg would be needed to cover 53 m^2 of wall?

9. A company need 33 workers to pack a its production in 25 days. If the total production needs to be packed in 15 days, how many <u>extra</u> workers do they need?

10. If I ride my bicycle at an average speed of 15 km/h, I travel a distance of 22 km in a certain period of time. If the speed is 17 km/h, how far will I travel?

11. A tap opened 9 hours during 8 days has thrown 5400 liters of water. How many liters will throw during 18 days, 8 hours per day?

12. If 18 lorries carry 1200 containers in 12 days, How long will 24 lorries need to carry 1600 containers?

13. We have used 2 cans with 1 Kg of paint, to paint a wall with dimensions 8 m x 2.5 m. How many cans, with 5 kg of paint, will we need to paint another wall with dimensions 50 m x 2 m?

Percentages

14. The population of a town is 652 000 and 35% of them live in the central district. How many of them live in this district?

15. In a sale the price of a television set is 150 € which is 75 % of the usual price, what was the original price?

16. The 6% of the population of Murcia are immigrants and there are 9900 immigrants living in this city, what is the population of Murcia?

17. The price of some clothes is 68 € and there is a discount of 7%, what is the final price?

18. I have bought a pair of jeans for 33 €, the IVA is 10 %, what was the price before?

19. The price of an electric oven before taxes is 560 € plus 17% IVA and the salesman offers a 12% discount, what is the final price?

20. Last year there were 1560 employees in a company, this year 312 new people have been employed. What has been the % increase of the staff in the company?

21. I have paid 161 € for a coat and the original price was 230 €. What is the % discount?

22. In an A shop one mobile costs 99 € plus IVA (16%). The same model costs in a B shop 114.90 € (IVA included). In what shop is the mobile cheaper?

23. At a sporting event that brings together 750 athletes, 30% of them are Americans, 8% Asians, 16% Africans, and the remainder Europeans. How many Europeans athletes take part in the match?

24. What's the number of guests who attend to a wedding, knowing that there are 33 men and 45% are women?

25. The price of a book after the increase of 20 % is 4.20 €. How much did it cost before the rise?

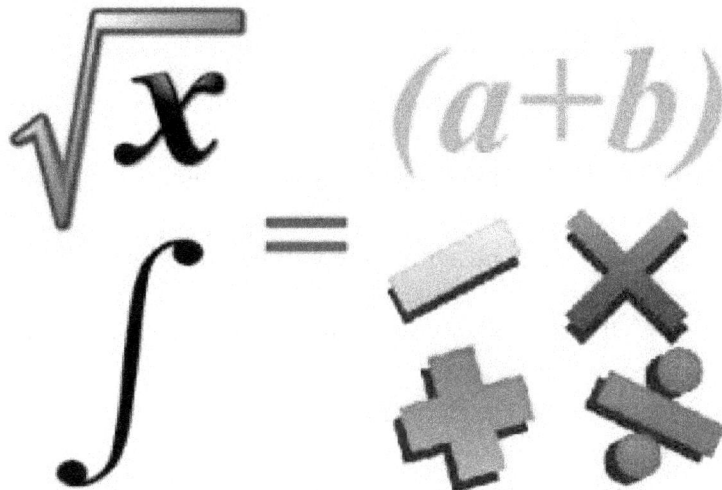

Unit 8.- Algebra and equations

1. Variables and expressions

A *variable* is a symbol that represents a number. We usually use letters such as x, n, p, t for variables.

Letters are useful if we want to operate with an unknown number instead with a particular one. Let us look at some examples:

- We say that *s* represents the side of a square, then *s* represents a number and:

4s Is the perimeter of the square $\qquad$ $s \times s = s^2$ Is the area of the square

When letters express numbers they have the same operating properties. The part of mathematics that deals with the study of the expressions with letters and numbers is called **algebra**.

An *expression* is a mathematical statement with numbers and variables.

For example, these are expressions: $\quad 3x \quad\quad x+3 \quad\quad 2(x-5) \quad\quad x \quad\quad 3x-1$

If Mark weights 80 kg and he gains n kg, the new weight is $80 + n$

131

1. Calling 'a' the age of a person today, write an expression for:

a) The age he/she will be in 5 years.

b) The age he/she was 7 years ago.

c) The age he/she will be after living the same time again.

2. Calling x a number, join each sentence with its corresponding expression:

a) The double of the number

b) The triple of the number plus eight units

c) The half of the number

d) The half of the number minus ten units

$$\frac{x}{2} - 10 \qquad \frac{x}{2} \qquad 3x + 8 \qquad 2x$$

3. Calling x a number, express in algebra:

a) The sum of the number and 10

b) The difference between 123 and the number

c) The double of the number

d) The triple of the number plus three units

e) The half of the number minus seven

f) The three quarters of the number plus forty-six

4. Join each sentence and its corresponding expression:

a) The distance moved in x hours by a car moving with speed 60 km/h.

b) The cost of x kg of oranges, which prices are 0.80 €/kg.

c) The area of a triangle having a base of 0.80 m and a height of x metres.

d) The age of Pedro, if his grandfather is now x years old, and he was 60 when Pedro was born.

$$\boxed{0,8x} \qquad \boxed{60x} \qquad \boxed{x - 60} \qquad \boxed{\frac{0,8 \cdot x}{2}}$$

5. Read and complete:

* Pablo´s salary is x euros.

* The gerent of the firm earns a double quantity than Pablo.

* The engineer earns 400 € less than the gerent.

* The person who cleans the office earns 10% less than Pablo.

Pablo	Gerent	Engineer	Cleaning service

6. Copy and complete:

n	1	2	3	4	5	10	100
5n – 3							

n	1	2	3	4	5	8	11
$\dfrac{2n-1}{3}$							

2. Monomials and operations

A monomial consists of the product of a known number **(coefficient)** by one or several letters with exponents that must be constant and positive whole numbers **(literal part)**.

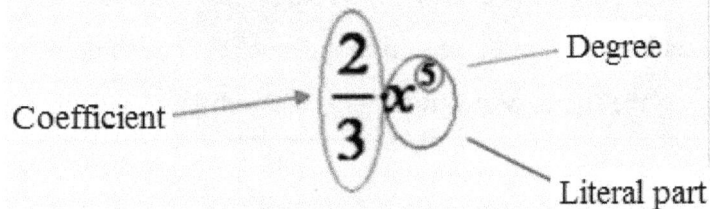

Coefficient $\dfrac{2}{3}x^5$ — Degree / Literal part

Example 1:

   a) 2x is a monomial. 2 is the coefficient, x is literal part.

   b) $-3x^2$ is a monomial, (-3) is the coefficient, x^2 is the literal part, x is the variable and degree 2.

   c) $\frac{3}{5}t^7$ is a monomial, $\frac{3}{5}$ is the coefficient, t^7 is the literal part, t is the variable and degree 7.

   d) $5xy^2$ is a monomial, 5 is the coefficient, xy^2 is the literal part, x and y are the variables and degree 3.

   e) 2x + 7 is an algebraic expression but it is not a monomial.

Exercises

7. Write the coefficient, literal part and degree of the following monomials:

Monomials	Coefficient	Literal part	Degree	Monomials	Coefficient	Literal part	Degree
$5x^3$				x			
$7x^4$				y			
$3y^6$				23			
$-4z^2$				$-8x^6$			
x^5				2t			
$-7y^3$				$5y^{12}$			
4z				h^7			
2				$2a^{10}$			

Two or more monomials are said to be **like monomials** if they have the same literal part.

2.1. Addition and subtraction of monomials

Two monomials can only be added or subtracted if they are **like monomials**. So, they must have the same literal part. In this case, we add or subtract the coefficients and we **leave the literal part unchanged**. When the literal part is different, the addition is left indicated.

Example 2: Calculate the following additions and subtractions of monomials:

a) $2x + 7x = 9x$ b) $15x + 2x = 17x$ c) $8x - 2x = 6x$

d) $x + 2x = 3x$ e) $5x^3 + 2x^3 = 7x^3$

f) $15x + 2 =$ we leave it indicated, because they are not like monomials.

Exercises

8. Simplify:

a) $x + x + x$ b) $a + a$ c) $2x - x$ d) $5a + 2a$ e) $3x + x$

f) $8a - 5a$ g) $4x - 3x$ h) $4a + 5a$ i) $7x - 7x$ j) $-3a + 4a$

k) $2x - 3x$ l) $3a - 7a$

9. Simplify:

a) $3x + 2x + x$ b) $10x - 6x + 2x$ c) $5a - 7a + 3a$

d) $a - 5a + 2a$ e) $-2x + 9x - x$ f) $-5x - 2x + 4x$

10. Simplify as much as possible:

a) $x + x + y$ b) $2x - y - x$ c) $5a + b - 3a + b$ d) $3a + 2b + a - 3b$

e) $2 + 3x + 3$ f) $5 + x - 4$ g) $2x - 5 + x$ h) $3x + 4 - 4x$

i) $x - 2y + 3y + x$ j) $2x + y - x - 2y$

11. Simplify as much as possible:

a) $x^2 + 2x^2$ b) $x^2 + x$ c) $3a^2 - a - 2a^2$ d) $a^2 - a - 1$

e) $x^2 - 5x + 2x$ f) $4 + 2a^2 - 5$ g) $2a^2 + a - a^2 - 3a + 1$ h) $a^2 + a - 7 + 2a + 5$

2.2. Product and division of two monomials

To **multiply** two or more monomials, we must multiply, <u>separately</u>, their coefficients and literal parts.

Remember the method to multiply powers having the same base: $x^m \cdot x^n = x^{(m+n)}$.

Example 3: Multiply: a) $5x^2 \cdot 7x^3$ b) $10x^4 \cdot 2x$ c) $\dfrac{3}{2}x^2 \cdot 4x^3$

Solution:

a) $5x^2 \cdot 7x^3 = \left\{\begin{array}{l} 5 \cdot 7 = 35 \\ x^2 \cdot x^3 = x^{(2+3)} = x^5 \end{array}\right\} = \boxed{35x^5}$ b) $10x^4 \cdot 2x = \left\{\begin{array}{l} 10 \cdot 2 = 20 \\ x^4 \cdot x = x^{(4+1)} = x^5 \end{array}\right\} = \boxed{20x^5}$

c) $\dfrac{3}{2}x^2 \cdot 4x^3 = \left\{\begin{array}{l} \dfrac{3}{2} \cdot 4 = \dfrac{3}{2} \cdot \dfrac{4}{1} = \dfrac{12}{2} = 6 \\ x^2 \cdot x^3 = x^{(2+3)} = x^5 \end{array}\right\} = \boxed{6x^5}$

To **divide** two or more monomials, we must divide, <u>separately</u>, their coefficients and literal parts.
Remember the method to divide powers having the same base: $x^m : x^n = x^{(m-n)}$.

Example 4: Divide: a) $15x^6 : 3x^2$ b) $8x^5 : 2x^3$ c) $10x^4 : 3x$ d) $\dfrac{3}{2}x^7 : \dfrac{9}{8}x^3$

Solution:

a) $15x^6 : 3x^2 = \left\{\begin{array}{l} 15 : 3 = 5 \\ x^6 : x^2 = x^{(6-2)} = x^4 \end{array}\right\} = \boxed{5x^4}$ b) $8x^5 : 2x^3 = \left\{\begin{array}{l} 8 : 2 = 4 \\ x^5 : x^3 = x^{(5-3)} = x^2 \end{array}\right\} = \boxed{4x^2}$

c) $10x^4 : 3x = \left\{\begin{array}{l} 10 : 3 = \dfrac{10}{3} \\ x^4 : x = x^{(4-1)} = x^3 \end{array}\right\} = \boxed{\dfrac{10}{3}x^3}$ d) $\dfrac{3}{2}x^7 : \dfrac{9}{8}x^3 = \left\{\begin{array}{l} \dfrac{3}{2} : \dfrac{9}{8} = \dfrac{3}{2} \cdot \dfrac{8}{9} = \dfrac{3 \cdot 2 \cdot 2 \cdot 2}{2 \cdot 3 \cdot 3} = \dfrac{2 \cdot 2}{2} = \dfrac{4}{3} \\ x^7 : x^3 = x^{(7-3)} = x^4 \end{array}\right\} = \boxed{\dfrac{4}{3}x^4}$

Exercises

12. Multiply:

a) $2 \cdot 5a$ b) $(-4) \cdot 3x$ c) $(-2a) \cdot a^2$ d) $5x \cdot (-x)$ e) $2a \cdot 3a$

f) $(-2x) \cdot (-3x^2)$ g) $2a \cdot (-5ab)$ h) $6a \cdot b$ i) $x \cdot 3x$

13. Calculate:

a) $3x^2 \cdot (-5x^4)$ b) $(-6x^3) \cdot (-7x)$ c) $(-5x^3) \cdot 2x^5$ d) $6x^2 \cdot 3x \cdot 2x^3$

14. Work out:

a) $3x \cdot 2x$ b) $(-6x) \cdot 3x^2$ c) $4x^3 \cdot (-2x)$ d) $4x^2 \cdot 3x^3$

e) $(-6x^3) : 2x$ f) $15x^5 : (-3x^4)$ g) $(-20x^3) : (-5x^3)$ h) $30x^7 : 5x^4$

A *polynomial* is the addition of two or more monomials.   For example: $3x^5 + 8x^3 - 3x + 5$.

The *degree* of a polynomial is the highest degree of its monomials.

3.1. Evaluation of a polynomial

To evaluate an expression at a number means we replace a variable in the expression with the number and then, if necessary, we calculate the value.

Example 5: Evaluate $5x + 3$ when $x = 2$.

<u>Solution:</u> $5x + 3$ becomes $5 \cdot 2 + 3 = 10 + 3 = \boxed{13}$.

---

15. Evaluate the following expressions:

a) $-\dfrac{1}{5}x^2$ when $x = 3$ b) $2t^2$ when $t = 2$ c) $a + b$ when $a = b = 7$

d) a^7 when $a = (-1)$ e) $n^4 + 3n$ when $n = 3$ f) $\dfrac{3n + 2}{5n - 3}$ when $n = 5$

16. Evaluate the polynomial $x^2 + 3x$,

a) when $x = 1$ b) when $x = 2$ c) when $x = 0$

17. Evaluate the polynomial $3x^2 - x + 3$,

a) when $x = 2$ b) when $x = 0$ c) when $x = (-2)$

18. Evaluate the polynomial $P(x) = 2x^3 + 2x - 3$

a) when $x = 1$ b) when $x = 2$ c) when $x = 0$ d) when $x = (-1)$

19. Evaluate the polynomial $P(x) = x^3 - 6x^2 + 8$

a) when $x = 0$ b) when $x = (-1)$ c) when $x = 2$.

Exercises

3.2. Expansion of brackets.

Multiplying a number by an addition is equal to multiplying by each adding number and then to adding the partial products.

Example 6: Expand the following expressions in brackets:

a) $2(x + 6)$　　　b) $3(2x - 7)$　　　c) $3x(3x+2)$

Solution:

a) $2(x + 6) = 2x + 12$

b) $3(2x - 7) = 6x - 21$

c) $3(2x - 7) = 6x - 21$

Exercises

20. Copy and fill in the missing terms:

a) $5(3x - 4) = \boxed{\dots} - 20$　　　　b) $x(x - 2) = \boxed{\dots} - 2x$

c) $a(4 - a) = \boxed{\dots} - a^2$　　　　d) $7x(3x - y) = \boxed{\dots} - 7xy$

21. Expand:

a) $2(x - 4)$　　　b) $6(2 - x)$　　　c) $6x(2x - 1)$　　　d) $2t(2t + 8)$

e) $5x(2x - 3y)$　　f) $x(x - 1)$　　　g) $2(7 + 7x)$　　　h) $2x(3x - a)$

22. Expand:

a) $x \cdot (x + 3)$　　b) $x \cdot (3x - 5)$　　c) $x^2 \cdot (4x^2 + 7x - 2)$　　d) $7 \cdot (5x^2 - 3)$

e) $3x \cdot (2x + 5)$　　f) $x \cdot (7x - 1)$　　g) $2 \cdot (x^2 - 2x + 1)$　　h) $2x \cdot (5x + 1 - 3x^2)$

23. Expand:

a) $5 \cdot (1 + x)$　　　b) $(-4) \cdot (2 - 3a)$　　　c) $3a \cdot (1 + 2a)$

d) $x^2 \cdot (2x - 3)$　　e) $x^2 \cdot (x + x^2)$　　　f) $2a \cdot (a^2 - a)$

24. Expand and simplify as much as possible:

a) $x + 2(x + 3)$　　　b) $7x - 3(2x - 1)$　　　c) $4 \cdot (a + 2) - 8$

d) $3 \cdot (2a - 1) - 5a$　　e) $2(x + 1) + 3(x - 1)$　　f) $5 \cdot (2x - 3) - 4 \cdot (x - 4)$

3.3. Factorization of expressions

It is important to know how to write expressions including brackets when it is possible; this is called **factorization** or **factoring**

Example 7: Factorize: a) 12x + 6   b) 21x − 14   c) $10x^2$ - 15x

Solution:

a) 12x + 6 = [2]·2·[3]·x + [2]·[3]·1 =   We mark the common factors   [2]·[3]· (2x + 1) = 6(2x + 1).

b) 21x − 14 = 3·[7]·x − 2·[7] = 7(3x − 2).

c) $10x^2$ - 15x = 2·[5]·[x]·x − 3·[5]·[x] = 5x(2x − 3)

We can check our answer by expanding the expression. *DO IT BY YOURSELF!!!*

Exercises	
	25. Factorize: a) 10x + 15 b) 4x −12 c) 9x − 9 d) $3x^2 − 2x$ e) $9x^2 + 3x$ f) 3x − 3 g) $33x^2 − 3x$ h) $5x − 3x^2$ i) $13x^2 − 2$ **26.** Factorize: a) $x^2 + 3x$ b) $3x^2 - 5x$ c) $4x^4 + 7x^3 - 2x^2$ d) $35x^2 - 21$ e) $6x^2 + 15x$ f) $7x^2 - x$ g) $2x^2 - 4x + 2$ h) $- 6x^3 + 10x^2 +2x$ **27.** Factorize, if possible: a) $-5x^2 + 2x^3$ b) $3x^2 - 9x^3$ c) $3x^2 - 3x$ d) $x^3 − x^2$ e) 7x - 4y f) $3x^2 + 2$ g) 12x - 4y h) $5x^2 - 1$

An *equation* is a statement in which two expressions are equal.

The letter in an equation is called the ***unknown***. (Sometimes it is called the ***variable***).

These are examples of equations:

$$x = 2 \qquad 3 = x \qquad 2x = 4 \qquad 5x+1 = 3 \qquad x^2 = 4$$

In an equation we consider two ***members***:

- First member is the member to the left of the sign "=".

- Second member is the member to the right of the sign "=".

Each monomial is called a term. Monomials with the same literal part are called ***"like terms"***

Example 8:

In the equation $3t - 2 = -7$ we say:

- The first member is $3t - 2$

- The second member is $- 7$

- There are three terms: $3t$; $(- 2)$ and (-7)

- The unknown is t.

In the equation $3x^2 - 6 = 5x + 2$, we say:

- The first member is $3x^2 - 6$

- The second member is $5x + 2$

- There are four terms: $3x^2$; $(- 6)$; $5x$ and 2

- The unknown is x.

4.1. Solution of an equation

The ***solution*** of an equation is a number that when substituted by the unknown, the equality is verified.

Example 9:

For the equation $x + 2 = 10$, the solution is $x = 8$, because $8 + 2 = 10$.

For the equation $3x - 5 = 1$, the solution is $x = 2$, because $3 \cdot 2 - 5 = 1 \rightarrow 6 - 5 = 1$.

For the equation $\dfrac{x - 2}{3} = 5$, the solution is $x = 17$, because $\dfrac{17 - 2}{3} = 5 \rightarrow \dfrac{15}{3} = 5$.

28. Check the given solutions for the following equations:

a) The solution for $5x + 2 = 22$ is $x = 4$

b) The solution for $\dfrac{x + 5}{2} = 4$ is $x = 3$

29. Join each equation with its solution:

a) $2x - 1 = 5$ $x = 10$

b) $x - 8 = 2$ $x = (-3)$

c) $2 = x - 5$ $x = 3$

d) $4 + x = 4$ $x = 0$

e) $3 - 3x = 6$ $x = (-1)$

f) $x + 4 = 1$ $x = 7$

---

5. Solving equations

5.1. Starting with equations

Solving an equation is finding out its solution. This means that we must have the unknown alone in one of the terms of the equation. We are seeing it with examples.

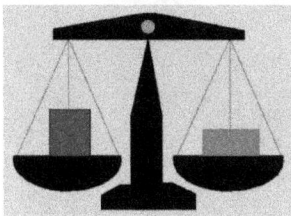 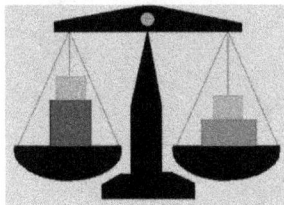

Figure 1   Figure 2

Look at the following equilibrated balance, in figure 1.

As you can see, you can add identical weights in both sides and the balance keeps equilibrated.

An equation can be considered as an equilibrated balance. So, we are going to be able to add or subtract identical *weights*.

Example 10: Solve the following equations:

a) x + 5 = 12

Solution: If we want to have the "x" alone, the number "5" must disappear. What would you do? If we subtract 5 in the first member, the number "5" will disappear. But if we subtract 5 in the first member, we must also subtract 5 in the second one:

$x + \cancel{5} - \cancel{5} = 12 - 5 \quad \rightarrow \quad x = 7$

CHECK: x + 5 = 12 → 7 + 5 = 12 *OK!!!*

b) x – 2 = 4

Solution: If we want to have the "x" alone, the number "(- 2)" must disappear. What would you do? If we add 2 in the first member, the number "(-2)" will disappear. But if we add 2 in the first member, we must also add 2 in the second one:

$x - \cancel{2} + \cancel{2} = 4 + 2 \quad \rightarrow \quad x = 6$

CHECK: x – 2 = 4 → 6 – 2 = 4 *OK!!!*

c) 3x = 12

Solution: If we want to have the "x" alone, the number "3" must disappear. What would you do? If we divide the first member by 3, the number "3" will disappear. But if we divide the first member by 3, we must also divide the second one by 3:

$\dfrac{\cancel{3}x}{\cancel{3}} = \dfrac{12}{3} \quad \rightarrow \quad x = 4$

CHECK: 3x = 12 → 3 · 4 = 12   *OK!!!*

d) $\dfrac{x}{4} = 7$

Solution: If we want to have the "x" alone, the number "4" must disappear. What would you do? If we multiply the first member by 4, the number "4" will disappear. But if we multiply the first member by 4, we must also multiply the second one by 4:

$\dfrac{x}{\cancel{4}} \cdot \cancel{4} = 7 \cdot 4 \quad \rightarrow \quad x = 28$ 　　　　 CHECK: $\dfrac{x}{4} = 7 \rightarrow \dfrac{28}{4} = 7 \rightarrow$ 　　　 *OK!!!*

e) $3 - x = 1$

Solution: If we want to have the "x" alone, the number "3" must disappear. What would you do? If we subtract 3 in the first member, the number "3" will disappear. But if we subtract 3 in the first member, we must also subtract 3 in the second one:

$3 - 3 - x = 1 - 3 \quad \rightarrow \quad - x = - 2 \quad \rightarrow$ But we do not want to have "- x", but "x". So, we must change the sign of the first member. Of course, we must also change the sign of the second member $\rightarrow x = 2$.

CHECK: $3 - x = 1 \rightarrow 3 - 2 = 1$ *OK!!!*

f) $- 3x = - 12$

Solution: If we want to have the "x" alone, the number "(-3)" must disappear. *BE CAREFUL WITH THE SIGNS* What would you do? If we divide the first member by (-3), the number "(-3)" will disappear. But if we divide the first member by (-3), we must also divide the second one by (-3):

$\dfrac{-3x}{-3} = \dfrac{-12}{-3} \quad \rightarrow \quad x = \dfrac{-12}{-3} = 4$

CHECK: $- 3x = - 12 \rightarrow - 3 \cdot 4 = - 12 \quad$ *OK!!!*

g) $5 = 2 - x$

Solution: Till now, we have left the "x" alone in the first member. But, if it is more comfortable, we can leave it in the second member. $5 = 2 - x$ In this case, if we want to have the "x" alone, the number "2" must disappear. What would you do? If we subtract 2 in the second member, the number "2" will disappear. But if we subtract 2 in the second member, we must also subtract 2 in the first one:

$5 - 2 = 2 - 2 - x \quad \rightarrow \quad 5 - 2 = - x \quad \rightarrow \quad 3 = - x \quad$ But we do not want to have "- x", but "x". So, we must change the sign of the second member. Of course, we must also change the sign of the first member $\rightarrow - 3 = x$.

CHECK: $5 = 2 - (-3) \rightarrow 5 = 2 + 3$ *OK!!!*

h) $8 = 2 - x$

Solution: Another option, if we do not like to have "-x", we can add "x" in both members.

$8 + x = 2 - \cancel{x} + \cancel{x}$ $\rightarrow$ $8 + x = 2$

Now, we must subtract "8" in both members. $8 - 8 + x = 2 - 8$ $\rightarrow$ $x = 2 - 8 = (-6)$

CHECK: $8 = 2 - x$ $\rightarrow$ $8 = 2 - (-6)$ $\rightarrow$ $8 = 2 + 6 = 8$ $OK!!!$

---

30. Solve the following equations:

a) $x + 9 = 14$ b) $x + 3 = 6$ c) $-13 + x = 9$ d) $x - 7 = -9$

e) $-35 = x + 22$ f) $3x = -27$ g) $5x = 20$ h) $12 = 4x$

31. Solve the following equations:

a) $9x = 45$ b) $110 = 10x$ c) $3x = 33$ d) $50 = 10x$

e) $75 = 25x$ f) $6 - x = 1$ g) $735 - x = 600$ h) $2x = -10$

(Exercises)

5.2. More difficult equations

In some cases, the equation has more than one term containing unknowns. For example,

$$4 + 2x = 6x - 10 + 2$$

Now, we are going to solve this kind of equations, step by step, with some examples:

Example 11: Solve the equation $4 + 2x = 6x - 10 + 2$

First of all, we must have together all the terms containing "x" in the same member. When you are not sure, take all the terms with "x" to the left member. You must do it as in the previous examples.

Step 1: "6x" term will disappear in the right member if we subtract "6x" in this member (and, of course, in the other one).

$$4 + 2x - 6x = \cancel{6x} - \cancel{6x} - 10 + 2$$
$$4 + 2x - 6x = -10 + 2$$
$$4 - 4x = -8$$

Step 2: Now, we are going to leave the term with "x" alone. We must remove the number "4". We must subtract 4 in both members.

$$\cancel{4} - \cancel{4} - 4x = -8 - 4$$
$$-4x = -12$$

Step 3: Now, we are going to leave the "x" alone. We must divide both members by (-4).

$$\frac{\cancel{-4}x}{\cancel{-4}} = \frac{-12}{-4} \qquad \rightarrow \qquad x = \frac{-12}{-4} = \boxed{3}.$$

Step 4: CHECK:
$$4 + 2x = 6x - 10 + 2 \qquad \rightarrow \qquad x = 4$$
$$4 + 2 \cdot 3 = 6 \cdot 3 - 10 + 2$$
$$4 + 6 = 18 - 10 + 2$$
$$10 = 20 - 10 \qquad\qquad OK!!!$$

Example 12: Solve the equation $\ \ 5 - 2x = 7x - 10 - 4x$

First of all, we must have together all the terms containing "x" in the same member. You are free to choose the member in which you take them. In this case, as there are more terms with "x" in the right member, we are going to take the "2x" to the right member.

Step 1: "(-2x)" term will disappear in the left member if we add "2x" in this member (and, of course, in the other one).

$$5 - \cancel{2x} + \cancel{2x} = 7x - 10 - 4x + 2x$$
$$5 = 7x - 10 - 4x + 2x \qquad \text{We add the "x" terms:}$$
$$5 = 5x - 10$$

Step 2: Now, we are going to leave the term with "x" alone. We must remove the number "(-10)". We must add 10 in both members.

$$5 + 10 = 5x \cancel{+10} \cancel{+10}$$
$$15 = 5x$$

145

Step 3: Now, we are going to leave the "x" alone. We must divide both members by 5.

$$\frac{15}{5} = \frac{5x}{5} \qquad \rightarrow \qquad \frac{15}{5} = x = \boxed{3}.$$

Step 4: CHECK:
$$5 - 2x = 7x - 10 - 4x$$
$$5 - 2 \cdot 3 = 7 \cdot 3 - 10 - 4 \cdot 3$$
$$5 - 6 = 21 - 10 - 12$$
$$-1 = 21 - 22 \qquad\qquad OK!!!$$

32. Solve the following equations:

a) $3x + 2 = 14$ b) $3 - 2x = 5$ c) $5x + 12 = 2$ d) $3 = 4 - 3x$

e) $2x = x + 3$ f) $5x - 2 = x + 1$ g) $2x - 3 = 2x + 1$ h) $3x + 1 = 7x - 1$

33. Solve the following equations:

a) $x + 8 + 2x = 6 - 2x$ b) $3 + 4x - 7 = x - 3$ c) $5x - 1 = 3x - 1 + 2x$

d) $6 - 3x + 2 = x + 7$ e) $2x + 5 - 3x = x + 19$ f) $7x - 2x = 2x + 1 + 3x$

34. Solve the following equations:

a) $11 + 2x = 6x - 3 + 3x$ b) $7 + 5x - 2 = x - 3 + 2x$ c) $x - 1 - 4x = 5 - 3x - 6$

d) $5x = 4 - 3x + 5 - x$ e) $3x - x + 7x + 12 = 3x + 9$ f) $6x - 7 - 4x = 2x - 11 - 5x$

g) $7x + 3 - 8x = 2x + 4 - 6x$ h) $5x - 7 + 2x = 3x - 3 + 4x - 5$

35. Solve the following equations:

a) $7x - 6 = x + 8 + 5x$ b) $5x + 32 = 4x + 41$ c) $2 + 3x + 2x = 4x + 9$

d) $4x - 5 + x = 5 + 3x - 1$ e) $6x + 2 - 4x = 9 - x + 8$ f) $10 + x + 14 = 30 + 5$

36. Solve the following equations:

a) $18 + 2x - 8 = x - 25$ b) $8x - 6 = x + 8 + 6x$ c) $4x - 12 + x = 4x - 1$

d) $5x + 4 = 20 + 2x$ e) $4 - 6x = 3 + 4x$ f) $5x - 2 = 7x + 12$

37. Solve the following equations:

a) $3x + 5 = 2x - 6$ b) $2x - 3 + 5x = -2 - 6x + 3$ c) $3x - 1 = 5 - 3x$

146

5.3. Equations with parentheses

In some cases, the equations have one or more parentheses. For example, $10 + 2(x - 3) = 4x - 8$

In these cases, first of all, we must **expand** the parentheses.

Now, we are going to solve this kind of equations, step by step, with some examples:

Example 13: Solve the equation $10 + 2(x - 3) = 4x - 8$

<u>Step 1</u>: First of all, we must expand the parentheses: $+ 2(x - 3) = + 2x - 6$ And include it in the equation:

$$10 + 2x - 6 = 4x - 8$$

<u>Step 2</u>: We must have together all the terms containing "x" in the same member. For example, we are going to take "4x" to the left member. "4x" term will disappear in the right member if we subtract "4x" in this member (and, of course, in the other one).

$$10 + 2x - 6 - 4x = 4x - 4x - 8$$

$$10 + 2x - 6 - 4x = -8 \qquad \text{Notice that } \begin{cases} 10 - 6 = 4 \\ 2x - 4x = -2x \end{cases}$$

$$4 - 2x = -8$$

<u>Step 3</u>: Now, we are going to leave the term with "x" alone. We must remove the number "4". We must subtract 4 in both members.

$$4 - 4 - 2x = -8 - 4$$

$$-2x = -12$$

<u>Step 4</u>: Now, we are going to leave the "x" alone. We must divide both members by (-2).

$$\frac{+2x}{+2} = \frac{-12}{-2} \qquad \rightarrow \qquad x = \frac{-12}{-2} = \boxed{6}.$$

<u>Step 5</u>: CHECK $10 + 2(x - 3) = 4x - 8 \qquad \rightarrow \quad x = 6$

$$10 + 2(6 - 3) = 4 \cdot 6 - 8$$

$$10 + 2(3) = 24 - 8$$

$$10 + 6 = 16 \qquad\qquad OK!!!$$

147

Example 14: Solve the equation $8 - 3(2 - 5x) = 4(x - 1)$

<u>Step 1</u>: First of all, we must expand the two parentheses: $\begin{cases} -3(2-5x) = -6+15x \\ 4(x-1) = 4x-4 \end{cases}$ And

include it in the equation:

$$8 - 6 + 15x = 4x - 4$$

<u>Step 2</u>: We must have together all the terms containing "x" in the same member. For example, we are going to take "4x" to the left member. "4x" term will disappear in the right member if we subtract "4x" in this member (and, of course, in the other one).

$$8 - 6 + 15x - 4x = 4x - 4x - 4$$

$$8 - 6 + 15x - 4x = -4 \qquad \text{Notice that } \begin{cases} 8 - 6 = 2 \\ 15x - 4x = 11x \end{cases}$$

$$2 + 11x = -4$$

<u>Step 3</u>: Now, we are going to leave the term with "x" alone. We must remove the number "2". We must subtract 2 in both members.

$$2 - 2 + 11x = -4 - 2$$

$$11x = -6$$

<u>Step 4</u>: Now, we are going to leave the "x" alone. We must divide both members by 11.

$$\frac{11x}{11} = \frac{-6}{11} \qquad \rightarrow \qquad x = \boxed{\frac{-6}{11}}$$

---

38. Solve the following equations:

a) $4 - (5x - 4) = 3x$

b) $7x + 10 = 5 - (2 - 6x)$

c) $5x - (4 - 2x) = 2 - 2x$

d) $1 - 6x = 4x - (3 - 2x)$

39. Find out the value of x:

a) $x - (3 - x) = 7 - (x - 2)$

b) $3x - (1 + 5x) = 9 - (2x + 7) - x$

c) $(2x - 5) - (5x + 1) = 8x - (2 + 7x)$

d) $9x + (x - 7) = (5x + 4) - (8 - 3x)$

Exercises

148

40. Solve the following equations:

    a) $2(x + 5) = 16$ b) $5 = 3 \cdot (1 - 2x)$ c) $5(x - 1) = 3x - 4$

    d) $5x - 3 = 3 - 2(x - 4)$ e) $10x - (4x - 1) = 5 \cdot (x - 1) + 7$

41. Find out the value of x:

    a) $6(x - 2) - x = 5(x - 1)$ b) $7(x - 1) - 4x - 4(x - 2) = 2$

    c) $3(3x - 2) - 7x = 6(2x - 1) - 10x$ d) $4x + 2(x + 3) = 2(x + 2)$

42. Solve the following equations:

    a) $3 (x + 6) = -2 (5 - x)$ b) $x + 9 = 2 (x - 6)$ c) $2x + 3 = 4x + 6 (x - 4) - 2$

    d) $1 + 4(x - 2) = -3x + 5(x + 1)$ e) $2(x + 6) - 7x = 3x$

5.4. Equations with fractions

In some cases, the equations include fractions. For example, $\dfrac{x + 2}{3} + \dfrac{5}{6} = 2x$

In these cases, first of all, we must transform each term in a fraction being **equivalent** with the originals.

Now, we are going to solve this kind of equations, step by step, with some examples:

Example 15: Solve the equation $\dfrac{x + 2}{3} + \dfrac{5}{2} = 2x$

Step 1: First of all, we must calculate the LCM (Lowest Common Multiple) of the denominators:

$$LCM(3, 2) = 6.$$

Step 2: Now, we are going to transform each term in a fraction being <u>equivalent</u> with the originals:

$$\frac{x+2}{3} = \frac{}{6}$$ Notice we have multiplied the denominator by 2. So, we also must multiply the numerator by 2: $2 \cdot (x + 2) = 2x + 4$ So, we have $\dfrac{x+2}{3} = \dfrac{2x+4}{6}$

$$\frac{5}{2} = \frac{}{6}$$ Notice we have multiplied the denominator by 3. So, we also must multiply the numerator by 3: $3 \cdot 5 = 15$ So, we have $\dfrac{5}{2} = \dfrac{15}{6}$

$$\frac{2x}{1} = \frac{}{6}$$ Notice we have multiplied the denominator by 6. So, we also must multiply the numerator by 6: $6 \cdot 2x = 12x$ So, we have $\dfrac{2x}{1} = \dfrac{12x}{6}$

Step 3: Now, we change each term by its equivalent fraction:

$$\frac{2x+4}{6} + \frac{15}{6} = \frac{12x}{6}$$

Step 4: As in previous equations, we can multiply both members by the same number, in this case, by 6 (this means, we can remove all the denominators in an equation, when they are all similar):

$$2x + 4 + 15 = 12x \quad \rightarrow \quad 2x + 19 = 12x$$

Step 5: We must have together all the terms containing "x" in the same member. For example, we are going to take "12x" to the left member. "12x" term will disappear in the right member if we subtract "12x" in this member (and, of course, in the other one).

$$2x + 19 - 12x = \cancel{12x} - \cancel{12x}$$
$$-10x + 19 = 0$$

Step 6: Now, we are going to leave the term with "x" alone. We must remove the number "19". We must subtract 19 in both members.

$$-10x+19-19=0-19$$

$$-10x=-19$$

Step 7: Now, we are going to leave the "x" alone. We must divide both members by (-10).

$$\frac{-10x}{-10}=\frac{-19}{-10} \quad \longrightarrow \quad x=\frac{-19}{-10}=\boxed{\frac{19}{10}}$$

Exercises	
	43. Solve the following equations:
	a) $\frac{3x+3}{2}=2x-7$ b) $x+7=\frac{x+5}{3}$ c) $\frac{9x+3}{2}=\frac{2x+4}{3}$
	d) $5-\frac{x+2}{3}=3-2x$ e) $2x+3=\frac{x+6}{3}$ f) $\frac{x+1}{3}-\frac{2x+2}{12}=1$
	g) $5x-\frac{3-2x}{2}=2x-\frac{5}{2}$ h) $\frac{15x}{2}-\frac{x}{4}=\frac{5}{2}$ i) $\frac{4x-6}{5}=\frac{x}{3}-\frac{4}{15}$
	j) $15x-\frac{6x-1}{2}-\frac{x-1}{3}=-6$

6. Problems with equations

Exercises	
	44. The double of a number minus five is seventeen. What is the number?
	45. The triple of a number plus four is forty-three. What is the number?
	46. Four times a number minus seven is 25. What is the number?
	47. Five times a number is the sum of the number plus 80. What is the number?
	48. Jennifer thinks of a number, doubles it and the answer is 18. What was the number?

49. Martin thinks of a number, subtracts 15 and the answer is 14. What was the number?

50. Claudette thinks of a number, doubles it and adds 4. The answer is 25. What was the number?

51. Jhosir thinks of a number, multiplies it by 3 and subtracts 5. The answer is 25. What was the number?

52. Sean thinks of a number, halves it and adds 8. The answer is 20. What was the number?

53. The perimeter of this rectangle is 40 cm. Find a.

a

6 cm

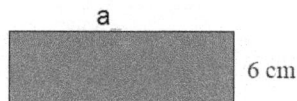

54. The area of this rectangle is 28 cm^2. Find d.

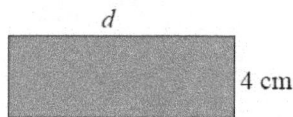

d

4 cm

55. The perimeter of this rectangle is 26 cm. Find n.

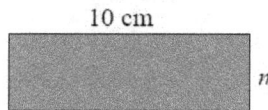

10 cm

n

56. The area of this rectangle is 20 cm^2. Find x.

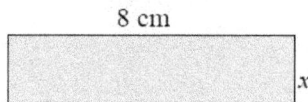

8 cm

x

57. The sum of my age and 7 is 42. Find my age.

58. The difference of a number and 23 is 124. Find the number.

59. The quotient of 35 and a number is 7. Find the number.

60. If someone gives me 24€ I will have 34.23 € Find the money I have.

61. The double of my age minus 7 years is the age of my elder brother who is 19 years old.

62. A teacher gives x coloured pencils each one of 8 girls except to one of them who only receives 5 pencils. The teacher gives 53 pencils in total. How many pencils did each girl receive?

63. Donald thinks of a number, multiplies it by 3 and subtracts 7. His answer is twice the number plus 5 units. Which is the number?

64. In a triangle, the smallest angle is 20° less than the median angle and the largest angle is twice the median one. Find all the angles.

65. If we shorten 2 cm each of the two opposite sides of a square, we get a rectangle with an area which has 28 cm^2 less than the area of the square. Find out the perimeter of the rectangle.

Review exercises

1. For the following monomials, indicate the coefficient, the literal part and the degree:

a) $2x^2$
b) $-3x^2y$
c) $3ac^3$
d) $-\dfrac{5}{7}xy$

2. Complete the following chart:

Monomials	Coefficient	Literal part	Degree
$6x^3$			
$-4x$			
Xy			
$-2a^2b$			

3. Write a monomial having:

a) As coefficient $\frac{1}{5}$ and as literal part, xy

b) As coefficient (-1) and as degree, 3.   **Sol.:** a) $\frac{1}{5}xy$ b) $-x^3$.

4. Calculate the following sums and subtractions of monomials:

a) 3x + 2x + 1 + 4   *b)* 6x - 3x + 7 – 2   *c)* 4x - 12x + 4 - 8

d) 4x - 3 - 12x – 8   *e)* $3x^2$ - 4x + 7x - $2x^3$ *f)* 3 - x +$3x^2$ +4x - x^2 +7

Sol.: a) 5x + 5; b) 3x + 5; c) -8x – 4; d) -8x – 11; e) $-2x^3 + 3x^2 + 3x$; f) $2x^2 + 3x + 10$.

5. Calculate the following products and quotients of monomials:

a) 3x · 2x   *b)* (- 6x) · $3x^2$ *c)* $4x^3$ · (-2x)   *d)* $4x^2$ · $3x^3$

e) $(-6x^3)$: 2x   *f)* $15x^5$: $(-3x^4)$ *g)* $(-20x^3)$: $(-5x^3)$ *h)* $30x^7$: $5x^4$

6. Evaluate the following expressions for the indicated values:

a) 2 · x + 3, when x = 1

b) 3 · a + 5 · b, when a = (-1) and b = 2

c) x^2 - 3·x, when x = (-1) and x = 2   **Sol.:** a) 5; b) 7; c) 4 and (-2).

7. Evaluate the monomial 6x - 3 when:

a) x = 1 *b)* x = 2   *c)* x = (-1)   *d)* x = (-3)   **Sol.:** a) 3; b) 9; c) (-9); d) (-21).

8. Evaluate the following expressions:

a) x^2 + 1 when x = 2   *b)* 3 - x^2 + 2x when x = 1   *c)* - $4x^2$ + 2 when x = (-2)

d) - 6x - 2 when x = (-3)   *e)* 2x + x^2 + 1 when x = 4   *f)* x^3 - 2x + 2 when x = (-3)

g) x + x^3 - x when x = (-1)   *h)* x^4 + 2 when x = (-1)

Sol.: a) 5; b) 4; c) (-14); d) 16; e) 25; f) (-19); g) (-1); h) 3.

9. Solve the following equations:

a) 2x - 7 = 3x - 8   b) 2x - 4 = 3x + 2   c) 2(x - 7) = -4(x - 1)

d) 3(6 + x) = 2(x - 6)   e) 2(5 + x) = 6x + 2   f) 9(x - 1) = 6(x + 3)

g) 8 - (4 - x) = 9   h) 12 - (x - 3) = 6   i) 7 - (2x - 3) = 2

Sol.: a) x=1; b) x=(-6); c) x=3; d) x=(-30); e) x=2; f) x=9 ; g) x=5 ; h) x=9 ; i) x=4.

10. Find out the value of x:

a) $2(x + 3) - 6(x + 5) = 3x + 4$

b) $-2x + 3(x - 1) = -12 + 5(2 - x)$

c) $5(x - 1) - 6x = 3(x - 3)$

d) $3(5x + 9) - 3(x - 7) = 11(x - 2)$

Sol.: a) x = (-4); b) $x = \dfrac{1}{6}$; c) x = 1 ; d) x = (-70) .

11. A farmer has chickens and horses. In his farm, there are 27 heads and 78 legs. How many chickens and horses does he have? **Sol.:** 15 chickens and 12 horses.

12. In a restaurant there are chairs (4 legs each) and stools (3 legs each). We have counted 44 seats and 164 legs. How many chairs and stools are there?

Sol.: 32 chairs and 12 stools.

13. Irene has found a bag with 14 coins, some of 20 cents and some of 10 cents. Their total value is 2 euros. How may coins of each value does she have?

Sol.: 8 coins of 10 cents and 6 coins of 20 cents.

16. If you multiply by 3 a number and you subtract 16 to the product, you get 29. What is the number? **Sol.:** 15.

17. If I add a number, its following and its previous, the result is 117. What is the number?

Sol.: 39.

18. The sum of three consecutive numbers is 84. What are these numbers? **Sol.:** 27, 28 and 29.

19. In my school, we are 624 students. If there are 36 more girls than boys, how many boys and girls are there? **Sol.:** 294 boys and 330 girls.

20. A fruit yoghurt is 5 cents more expensive than a natural one. For six fruit yogurts and four natural yogurts we have paid 4.80 €. How much do they cost?

Sol.: Natural: 45 cent; fruit: 50 cent.

21. At a shop, nails are sold in boxes of three different sizes: large, median and little. A large box contains twice the number of nails in the median, and this, double that small one. If you buy one box of each size, you get 350 nails. How many nails does each box contain?

Sol.: 50, 100 and 200.

22. The price for one kg of oranges is double than for pears. For three kg of oranges and four kg of pears I had to pay 11 €. What are their prices?

Sol.: 2.2 € and 1.1 €.

23. In an exam containing 50 test questions, they give you 3 points for each correct answer, but they subtract you 2 points for each wrong answer. You have answered all the questions and you have 85 points. How many answers were correct?

Sol.: 37.

24. A rectangular field is 18 meters longer than wide. It has a wall 156 meters long. What is the length and width of the field?

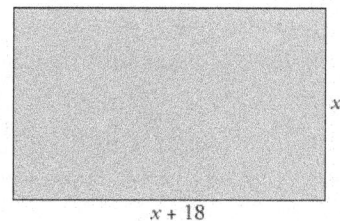

x

$x + 18$

Sol.: 30 m x 48 m.

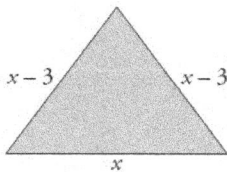

25. The two equal sides of a isosceles triangle are 3 cm shorter than the unequal one. Its perimeter is 48 cm. What are the lengths of its sides?

$x - 3$ $x - 3$

x

Sol.: 18, 15 and 15 cm.

Unit 9.- Statistics and probability

1. Statistics

Statistics have become an important part of everyday life. We are confronted by them in newspapers and magazines, on television and in general conversations. We encounter them when we discuss the cost of living, unemployment, medical breakthroughs, weather predictions, sports, politics and the state lottery. Although we are not always aware of it, each of us is an informal statistician. We are constantly gathering, organizing and analyzing information and using this data to make judgments and decisions that will affect our actions.

Population and sample

Statistics is a branch of mathematics in which groups of measurements or observations are studied. Statistics objectives are collecting, organizing and analyzing numerical facts.

Statistics work begins when collecting data. As you can imagine, if you want to study heights of citizens in a town, you cannot measure all the *population*. You must choose an appropriate *sample*.

A *population* is a complete set of items that is being studied. It includes all members of the set. The set may refer to people, objects or measures that have a common characteristic.

Examples of a population are all high school students, all cats, all citizens in a town.

A *sample* is relatively small group of items selected from a population. If every member of the population has an equal chance of being selected for the sample, it is called a *random sample*.

For example, the owner of a screw factory wants to make a quality control. He picks up 1 out of every 100 produced screws and then he analyses them.

- The *population* is the total number of the screws of the factory.
- The *sample* is 1% of the population.
- The *individuals* are each one of the screws.

2. Statistical variables

Data are numbers or measures that are collected. Data may include numbers of individuals that make up the census of a city, ages of pupils in a certain class, temperatures in a town during a given period of time, sales made by a company, or test scores.

Variables are characteristics or attributes that enable us to distinguish one individual from another. They take on different values when different individuals are observed.

Some variables are height, weight, age and price. Variables are the opposite of *constants* whose values never change.

Statistical variables can be classified in several ways:

Quantitative and qualitative variables

> - A **_quantitative or numerical variable_** is that one whose data are numbers. For example, heights, number of books, incomes in €, etc.
>
> - A **_qualitative or categorical variable_** is that one whose data have labels (i.e. words). For example, your favoutite musical group, a list of the products bought by different families at a grocery store, eyes colour, etc.

Exercises

1. Classify the following variable into quantitative or qualitative ones:

a) Favoutite sport   b) Shoe's number   c) Favourite future job

d) Maths mark in the latest exam   e) Number of books read in the last month.

2. For each of the following cases, indicate what are the population, the variable and the type of variable.

a) Weight of babies that were born last year in Dublin.

b) Favourite subject for the students in school.

c) Number of pets in French households.

d) Political party that the Spanish electors are going to vote for in the next local elections.

e) Weekly time that students from 12 to 16 spend on reading in Italy.

3. Frequency distributions. Frequency tables

Groups of data have little value until they have been placed in some kind of order. Usually measurements are arranged in ascending or descending order. Such a group is a *distribution*. A *frequency distribution* is a table in which measurements and the *frequency* or total number of times that each item occurs is recorded.

Example: The number of televisions in each house of my street is shown in the frequency table:

Number of TVs	Number of houses
0	1
1	5
2	12
3	9
4	1

a) Calculate the number of houses in my street.

b) Calculate the total of number of televisions in my street.

Solution:

a) The total number of houses is: $1 + 5 + 12 + 9 + 1 =$
= 28 houses.

b) The total number of televisions is: $(0 \cdot 1) + (1 \cdot 5) + (2 \cdot 12) + (3 \cdot 9) + (4 \cdot 1) = 0 + 5 + 24 + 27 + 4 =$
= 60 televisions.

Absolute and relative frequencies

The **absolute frequency** is the number of times a determined value of a variable occurs.
We will write f_i the absolute frequency of x_i.

We have to know that $f_1 + f_2 + ... + f_n = N$, where N is the total number of data. (*n* is the number of different data).

The **relative frequency** is the absolute frequency divided by the total number of observations or total number of data. We will write h_i the relative frequency of x_i.

And we have $h_1 + h_2 + ... + h_n = 1$

3. In the list below is shown the qualification in Maths of 20 students in a class:

$$\begin{array}{ccccc} 3 & 3 & 3 & 4 & 5 \\ 5 & 5 & 5 & 5 & 6 \\ 6 & 7 & 7 & 7 & 7 \\ 8 & 9 & 9 & 9 & 9 \end{array}$$

a) Build a table writing the absolute and relative frequencies of this set of data.

x_i	Absolute frequency, f_i	Relative frequency, h_i
3		
4		
5		
6		
7		
8		
9		

b) Calculate the sum of all the absolute frequencies and of all relative frequencies.

4. Mrs. Parker asked her students about their favourite subject in the school, and the answers were:

Maths	English	Sciences	English	Maths
History	English	Music	English	Maths
English	Maths	English	History	English
English	English	English	English	Maths
History	Sciences	Maths	Maths	Sciences

Make a table and show the absolute and relative frequencies of these data.

5. In thirty shots, a man makes the following scores:

5 2 2 3 4   4 3 2 0 3   0 3 2 1 5

1 3 1 5 5   2 4 0 0 4   5 4 4 5 5

Build a table and summarize the absolute and relative frequencies of the different scores.

Cumulative absolute and relative frequency

- **Cumulative Absolute Frequency** of a datum x_i is the sum of the absolute frequencies of values less or equal than it. We write F_i.

$$F_i = f_1 + f_2 + ... + f_i$$

- **Cumulative Relative Frequency** of a datum x_i is the sum of the relative frequencies of values less or equal than it. We write H_i. It is the quotient between the cumulative absolute frequency and the total number of data.

$$H_i = h_1 + h_2 + ... + h_i$$

6. It is shown the number of brothers and sisters of the 25 students:

02101   11213   01111   01123   21042

Complete the table with the different frequencies:

x_i	f_i	F_i	h_i	H_i
0				
1				
2				
3				
4				

7. We asked 20 students of 2nd E.S.O about the age of their fathers. The answers were the following:

40 42 44 47 44 41 43 46 42

46 44 45 47 42 45 40 43 47

Complete the table with the different frequencies:

x_i	f_i	F_i	h_i	H_i
40				
41				
42				
43				
44				
45				
46				
47				

4. Statistical graphs

Many people find it easier to obtain information from pictures than from written material. Statisticians display mathematical relationships with diagrams and graphs. From these pictures numerical data can be summarized clearly and easily.

Bar diagrams

When the data of a frequency distribution have not been grouped in intervals, they can be represented on a *bar diagram*. A bar diagram illustrates the pattern of a distribution. It clearly shows whether the data are spread out evenly or if they tend to cluster about any point.

To build a bar diagram, list the measurements, from lowest to highest, horizontally across the bottom of the graph. On the left side vertically list the frequencies or number of times that the measurements occur. Finally, on each measurement, draw a bar with height equal to its frequency. We draw all the bars with the same width.

Bar diagram is suitable for discrete variable (numerical or categorical).

For example, following figure shows a bar diagram indicating frequencies of vowels in a sentence in English:

Vowels in a sentence in English

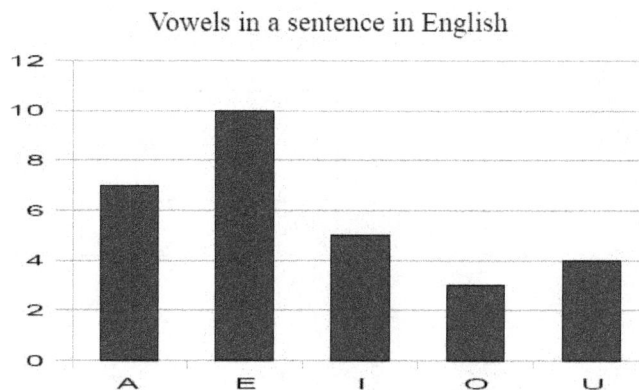

Histograms

The **histogram** is used for variables whose values are numerical and measured on an interval scale (usually for continuous variables).

A histogram divides up the range of possible values in data set into classes or groups. For each group, a rectangle is constructed with an area proportional to the frequency, (if the bars have equal width the height of each bar corresponds to the frequency).

Example: Mario decided to collect data about the height of his classmates in the school. These are the data of 40 children (in cm):

163	167	165	159	164	168	161	164	163	164
165	163	167	165	164	164	168	161	164	165
163	164	170	160	157	167	165	172	165	167
164	164	168	161	164	163	164	155	158	162

Solution:

It is useful to make a group frequency table for this case:

Interval	Tally	Frequency
[155-160)	////	4
[160-165)	### ### ### ### /	21
[165-170)	### ### ///	13
[170-175)	//	2

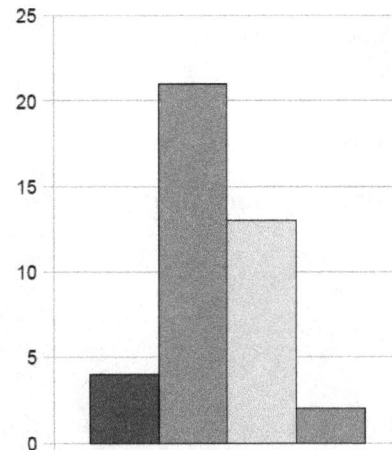

---

8. These are the heights, in cm, of 40 people.

153	134	155	142	140	163	150	135	170	156	171	161	141
153	144	163	140	160	172	157	136	160	134	154	176	154
173	179	160	152	170	148	151	165	138	143	147	144	156
139												

Complete the following frequency table and draw the histogram.

Height (cm)	Number of people
[130,140)	
[140,150)	
[150,160)	
[160,170)	
[170,180)	

Exercises

9. This frequency table shows the times for 50 runners in a Marathon.

Time (hours)	Number of runners
[1,2)	0
[2,3)	15
[3,4)	23
[4,5)	18
[5,6)	8

Draw a histogram to show the times.

Pie charts

A **pie chart** is a way of summarizing a set of categorical data. It is a circle which is divided into sectors. Each sector represents a particular category. The area of each sector is proportional to the number of cases in that category.

Example: We ask 240 people to name their favourite fruit. With these results, draw a pie chart to illustrate the information.

Fruit	Apple	Banana	Orange	Other
Number of people	50	80	72	38

Solution: First, we calculate the angle for one person: $360° : 240 = 1.5°$ → Each person corresponds an angle of 1.5°.

Then, we calculate the angles of each category:

- Apple: $50 \cdot 1.5° = 75°$
- Banana: $80 \cdot 1.5° = 120°$
- Orange: $72 \cdot 1.5° = 108°$
- Other: $38 \cdot 1.5° = 57°$

And then, we measure, colour and label the sectors:

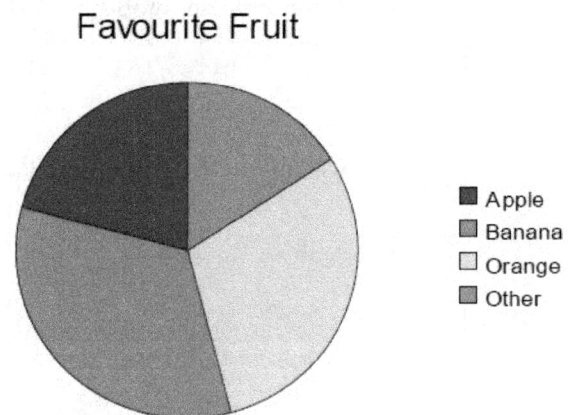

Favourite Fruit

- Apple
- Banana
- Orange
- Other

10. The Manchester United football team plays 36 matches in a season. They win 15 matches, and they draw just 8 matches.

a) Calculate the number of matches that they lose.

b) Calculate the angle one match represents in a pie chart.

c) Calculate the angle of each category in the pie chart.

d) Draw a pie chart to show the information.

11. The weather record for 60 days is shown in the frequency table.

Weather	Number of days
Sunny	15
Cloudy	18
Rainy	14
Snowy	3
Windy	10

a) Calculate the angle one day represents in a pie chart.

b) Calculate the angle of each category in the pie chart.

c) Draw a pie chart to show the data.

Frequency polygons

A **frequency polygon** is a line graph which can be used to represent the frequency of a set of numbers. It is formed by connecting a series of points. The abscissa of each point is the midpoint of the interval in which the point lies. The ordinate of each point is the frequency for the interval. The polygon is closed at each end by drawing a line from the endpoints to the horizontal axis at the midpoint of the next interval.

Example: In an intelligence test, 200 people have been examined. Results are given in the following table. Draw a histogram to represent them and build, also, the frequency polygon.

Interval	Mark
30 – 40	6
40 – 50	18
50 – 60	76
60 – 70	70
70 – 80	22
80 - 90	8

Solution:

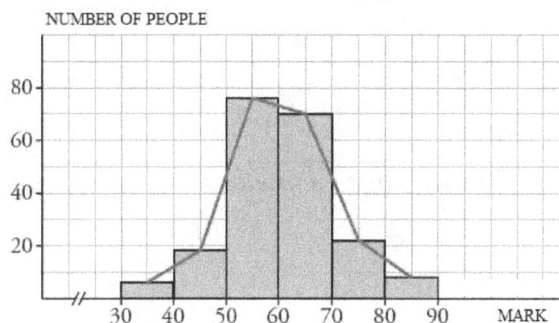

167

12. These are the heights, in cm, of 40 people.

153	134	155	142	140	163	150	135	170	156	171	161	141
153	144	163	140	160	172	157	136	160	134	154	176	154
173	179	160	152	170	148	151	165	138	143	147	144	156
139												

Complete the following frequency table and draw the frequency polygon:

Height (cm)	Number of people
[130,140)	
[140,150)	
[150,160)	
[160,170)	
[170,180)	

5. Measures of central tendency

A ***measure of central tendency*** is a single value that tries to describe a set of data by identifying the central position in that set of data. As such, measures of central tendency are sometimes called measures of central location. The *mean* (often called the average) is most likely the measure of central tendency that you are most familiar with, but there are others, such as, the *median* and the *mode*.

In the following sections we will look at the mean, mode and median and learn how to calculate them and under what conditions they are most appropriate to be used.

Mean

The ***average*** or ***mean*** of a list of numbers is the total of all values divided by the number of values.

To calculate the mean, we can use the absolute frequencies of values, multiplying every value by its absolute frequency, then adding all these products and finally dividing by the number of values. It is shown in the formula:

$$\bar{x} = \frac{x_1 \cdot f_1 + x_2 \cdot f_2 + \ldots + x_n \cdot f_n}{N} = \frac{\sum x_i \cdot f_i}{N}$$

Example: Find the mean of 10, 11, 7 and 8:   Solution: $\bar{x} = \frac{10+11+7+8}{4} = \frac{36}{4} = \boxed{9}$.

Example: Find the mean of the following values:

4 4 4 4 5 5 6 6 6 6

6 7 8 8 8 9 9 9 9 10

Solution: We can use a frequency table:

x_i	f_i
4	4
5	2
6	5
7	1
8	3
9	4
10	1

$$\bar{x} = \frac{4\cdot4 + 5\cdot2 + 6\cdot5 + 7\cdot1 + 8\cdot3 + 9\cdot4 + 10\cdot1}{20} = \frac{16+10+30+7+24+36+10}{20} = \frac{133}{20} = \boxed{6.65}.$$

Exercises

13. Calculate the mean of: 5, 3, 54, 93, 83, 22, 17 and 19.

14. Calculate the mean of values given as a table:

x_i	f_i
2	5
3	5
4	7
5	8

Median

The **median** of a list of values is found by ordering them from least to greatest. Once they are arranged, median is the central value. Median is the value that has as many values less than it as higher than it.

- If the list has an **odd** number of values, the middle number in this ordering is the median.
- If there is an **even** number of values, the median is the average of the two middle numbers.

Example: Find the median of the following sets of values: a) 8, 7, 5, 6, 9;   b) 9, 6, 7, 5, 6 ,7.

Solution:

a) First, we have to order the numbers: 5, 6, 7, 8, 9. As there are five (odd) data, median is 7. $\boxed{Me = 7}$.

b) We order the numbers: 5, 6, 6, 7, 7, 9. There are six (even) data, so, we get the two middle numbers, (6 and 7) and we calculate the mean of both numbers: $Me = \dfrac{6+7}{2} = 6.5$. Median is 6.5. $\boxed{Me = 6.5}$.

Mode

The **mode** in a list of numbers is the most repeated number (or numbers). It is the value with the greatest absolute frequency.

Example: Find the mode of following sets: a) 2, 3, 3, 6, 4, 3, 2, 5, 6, 3;

b) 2, 2, 2, 3, 4, 4, 4, 5, 6, 6.

Solution:

a) The mode is 3, because it is the number that appears most often in the list. $\boxed{Mo = 3}$.

b) We have two modes in this case, 2 and 4, because they appear three times. Both numbers have the greatest absolute frequency. $\boxed{Mo = 2, 4}$.

15. The tallest 4 trees in a park have heights in meters of 40, 52, 50 and 55. Find the mean and the median of their heights.

16. Find the mean, median and mode for the following data:

10, 12, 13, 12, 13, 10, 14 and 13.

17. Twenty families are asked about how many children they have. These are the answers:

3 3 4 1 2 3 2 5 1 0

2 2 3 2 4 2 5 3 4 3

a) Complete the table with the frequencies: f_i, h_i, F_i and H_i.

b) Find the mean, median and mode.

---

6. Introduction to probability. Experiments. Sample space. Events

Introduction

Humans often think in probabilistic terms, whether we are conscious of it or not. That is, we decide to cross the street when the probability of being run over by a car is sufficiently low, we go fishing at the lakes where the probability of catching something is sufficiently high, and so on. So, even when people are wholly unfamiliar with the

mathematical formalization of probability, there is an inclination to frame uncertain future events in such terms.

Deterministic and random experiments

In our life, in Nature, we can have two types of experiences or, in probabilistic terms, two types of experiments can be done: *deterministic* and *random* experiments. A **deterministic** experience or experiment is that one in which you can know the result before performing it. For example, if today it is Monday, we are *sure* tomorrow it will be Tuesday. Or, if I hold a pencil in my hand and I open my fingers, I am sure, even if I do not do it, pencil will fall, it won't go up. So, deterministic experiences are not in the field of study of probability, because they produce *sure* results.

On the other hand, **random** experiments refer to those experiments whose results cannot be determined before performing them. For example, if I throw a coin on a table, I can have two results, and I cannot determine it if I do not throw it. As we cannot be sure about the result of

random experiments, we speak about *probability* of a result to happen.

Probability is the branch of Mathematics that studies and quantifies the chance of random experiences results.

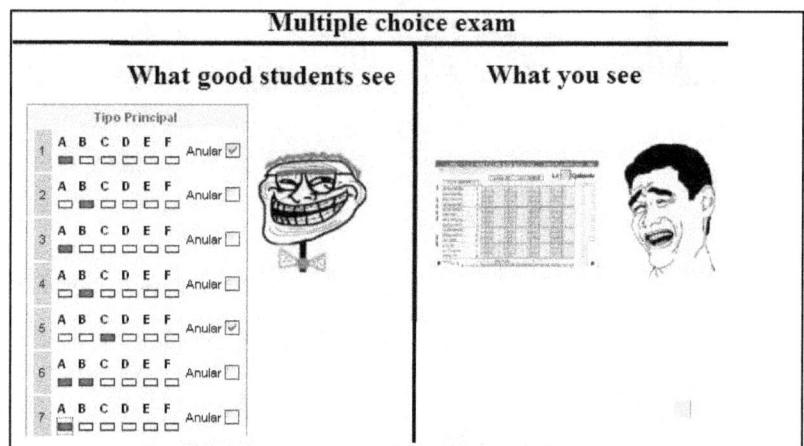

Sample set. Elements

Each possible simple result of a random experiment is named ***element***. The ***sample set*** of a random experiment includes all the elements in it.

Example: In a box, there are 3 green balls, 2 red ones and 6 blue. Write its sample set. How many elements does it have?

Solution: Sample set: {G, R, B}. Three elements.

Example: In a bag, there are 10 numbered cards, from 0 to 9. Write its sample set. How many elements does it have?

Solution: Sample set: {0, 1, 2, 3, 4, 5, 6, 7, 8, 9}. Ten elements.

Example: Two dice are thrown on a table. We study the sum of them. Write its sample set. How many elements does it have?

Solution: Sample set: {2, 3, 4, 5, 6, 7, 8, 9, 10, 11, 12}. Eleven elements.

"Dice" is the plural of *"die"*. So, in a game, you can throw one *die* or some *dice*.

Events

An *event* is a sub-set of the sample set of a random experiment.

Example: We throw a die on a table. As you already know, its sample set is {1, 2, 3, 4, 5, 6}. Describe the following events:

    A = Obtain an even number,   B = obtain a multiple of three,

    C = obtain a number higher than 3,   D = obtain a number equal or higher than 3,

    E = obtain a number between 1 and 6 , both included,   F = obtain number 8.

Solution: A = {2, 4, 6}   B = {3, 6}   C = {4, 5, 6}   D = {3, 4, 5, 6}

            E = {1, 2, 3, 4, 5, 6}   F = Ø (empty set).

There are several special events we must consider:

- *Sure event* is that one that is always going to happen. Its probability is 100%.

- *Impossible event* is that one that is never going to happen. Its probability is 0.

- Given an event A, we can define its *complementary event*, denoted by $\overline{A}$, as a new event that includes the elements of sample set not included in A. Notice that A + $\overline{A}$ = sample set.

Example: In a box, there are balls numbered from 1 to 10. Describe the following events:

a) A = Obtain a multiple of 4

b) B = Obtain a number from 1 to 10

c) C = Obtain number 20

d) D = Obtain a divisor of 8

e) $\overline{D}$

f) How could you define elements in $\overline{D}$?

Solution:

a) A = {4, 8}

b) B = {1, 2, 3, 4, 5, 6, 7, 8, 9, 10} (sure event)

c) C = Ø (impossible event).

d) D = {1, 2, 4, 8}

e) E = {3, 5, 6, 7, 9, 10}

f) Numbers from 1 to 10 not being divisors of 8.

Exercises

18. Classify the following experiments into deterministic and random experiments:

a) Extract a card from a cards game.

b) Check the weight of a liter of mercury.

c) Ask a number to a friend.

d) Throwing three dice and write down their sum.

e) Subtracting two known numbers.

19. Write all the possible results of throwing two coins on a table.

20. We throw a coin and a die. Write the sample set.

21. In a box, there are balls numbered from 1 to 20. Describe the following events:

a) A = Obtain a multiple of 5

b) B = Obtain number 18

c) C = Obtain a divisor of 40

d) $\overline{A}$

---

7. Laplace's Law

As we have already said, probability field is that one in which there are not sureness when performing an experience. So, we can only speak about the probability of an event to happen. In this part of the unit, we are going to learn to quantify this probability.

Expression to quantity probability was developed by Laplace. That is the reason by which it is known as **Laplace's Law.**

Probability of an event is calculated by counting favourable elements for the event and possible elements of the experiment, and calculating their quotient.

$$P(A) = \frac{Number\ of\ favourable\ elements\ to\ A}{Number\ of\ possible\ elements} \qquad \textbf{\textit{(Laplace's Law).}}$$

Example: In a bag, there are ten numbered balls from 1 to 10. A ball is extracted from the bag. Calculate the following possibilities:

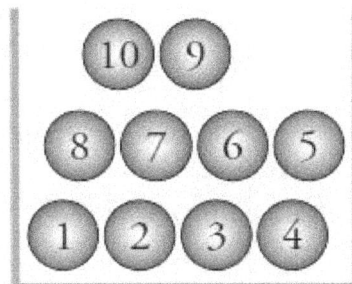

a) Obtaining number 8   b) Obtaining an even number

c) Obtaining a multiple of 5

<u>Solution:</u> In all cases, there are 10 possible elements, 10 balls.

a) There is 1 ball with number 8, so, $P(8) = \boxed{\dfrac{1}{10}}$

b) There are 5 even numbers between 1 and 10 (2, 4, 6, 8 and 10), so, $P(even) = \dfrac{5}{10} = \boxed{\dfrac{1}{2}}$

c) There are 2 multiples of 5 between 1 and 10 (5 and 10), so, $P(even) = \dfrac{2}{10} = \boxed{\dfrac{1}{5}}$

---

22. When throwing a die, calculate the probability of obtaining:

a) a multiple of 5   b) a divisor of 2   c) a prime number   d) Number 3

e) a divisor of 6   f) an even number divisor of 4   g) multiple of 7

h) less than 10   i) odd number

23. From a pack of Spanish cards we get a card.   a) What is the probability of getting a horse?   b) And a figure?   c) And a "gold"?

24. In a box there are 5 yellow balls and 7 red ones. What is the probability of getting a yellow one? And a green one?

Exercises

25. A card has a black side and white one. We have thrown this card 85 times and we have obtained the white side 43 times. What do you think probability of getting black one is?

26. We throw two dice and sum their punctuations. What is the probability of obtaining:

   a) 3   b) Higher than 10   c) 7   d) 4 or 5.

27. A box contains 4 white balls, 2 red ones and 5 black ones. Calculate the probability to obtain: a) a white ball   b) a red ball   c) a white or black ball.

28. In a bag, there are 90 balls, numbered from 1 to 90. We extract one of them.

   a) What is the probability of getting ball number 59?

   b) What is the probability of getting a multiple of 10?

29. In a bag there are balls having two sizes, large (L) and small (S). We extract one of them, write the size (L or S), insert it back to the bag, extract another one, etc. In this way, we observe 84 L balls and 36 S ones. What values would you give to P(L) and P(S)?

30. In a summer school, there are 32 European children, 13 American, 15 African and 23 Asian ones. Representative is randomly chosen. What is the probability of being an European one?

31. Two dice are thrown on a table and sum of punctuations are written. Calculate the probabilities of each sum.

Review exercises

1. Indicate, for each of the five cases proposed:
   • What is the population   • What is the variable.
   • Variable type: qualitative, quantitative, quantitative discrete or continuous.

   a) Birth weight of babies   b) Favourite professions for students c) Number of pets in houses
   d) Party which will be voted in the next general election.
   e) Weekly time spent reading by ESO students in Spain.

2. The percent of vehicles registered during the month of October 2006 is outlined in this table (data are approximate):

Type of vehicle	Percentage
Cars	69
Lorries	17
Motorbikes	
Busses	0.15
Others	1.45

a) Find the percent of motorcycles registered.

b) Calculate what was the total number of vehicles enrolled, knowing that exactly 279 busses were registered.

c) The set of vehicles registered, is population or sample?

d) Indicate what type of variable it is.

Sol.: a) 12.4%; b) 186 000; c) Population; d) Qualitative or categorical.

3. Figure 1 shows evolution of unemployed ratio in Spain from 1995 to 2000 year. Figure 2 shows a table for the same data, in which only difference is scale for Y-axis.

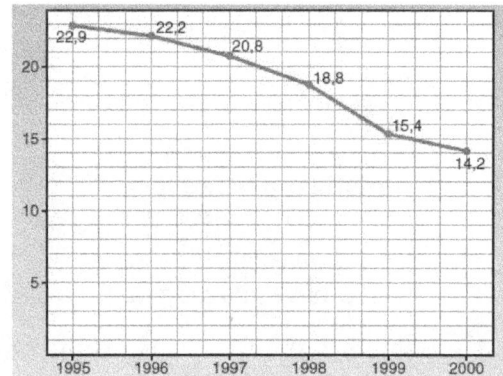

a) Do both of them give the same sensation?

b) Which one do you think a government would choose to give unemployment data and which one would be chosen by opposition party?

Tables and graphs elaboration

4. When being asked about the number of books they have read for last month, 3rd ESO students answers were the following:

$$2\ 1\ 3\ 1\ 1\ 5\ 1\ 2\ 4\ 3$$
$$1\ 0\ 2\ 4\ 1\ 0\ 2\ 1\ 2\ 1$$
$$3\ 2\ 2\ 1\ 2\ 3\ 1\ 2\ 0\ 2$$

a) Build a frequency table.

b) Draw the corresponding bar diagram.

5. Chosen colour by Spanish people when buying a car is shown in the following table. Draw the corresponding pie chart.

Colour	Percentage
Argent grey	36%
Black	22%
Blue	18%
Red	10%
White	8%
Green	4%
Others	2%

Statistical parameters

6. We have investigated, in different shops, price of a determined printer model, and we have obtained the following results, in euros:

$$146 - 150 - 141 - 143 - 139 - 144 - 133 - 153$$

a) Calculate the mean price.   b) What is the median?

Sol.: a) 143.625 €;   b) 143.5 €.

7. Counting the number of misprints in a book, Peter has obtained the following data:

Number of misprints	0	1	2	3	4	5
Number of pages	50	40	16	9	3	2

Calculate the a) mean, b) the mode.   **Sol.:** a) 1.008; b) 0 misprints.

8. Teresa and Rosa are basketball players. Their punctuations for a week training have been:

Teresa	16	25	20	24	22	29	18
Rosa	23	24	22	25	21	20	19

Calculate mean punctuation for each one.   **Sol.:** 22 and 22.

9. 40 people were asked about the number of people living at home:

4 5 3 6 3 5 4 6 3 2   2 4 6 3 5 3 4 4 5 3 6

4 5 7 4 6 2 3 4 4 3   4 4 5 3 2 6 3 7 4 3

a) Build a frequency table and draw the corresponding diagram.

b) Calculate the mean, median and mode.

Sol.: 4.125; 4 and 3.

10. In my high school there are 200 students. They were asked about their favorite pet. See the figure and answer the questions:

a) How many people did answer "dogs"?

b) And "cats"?   **Sol.:** a) 60; b) 50.

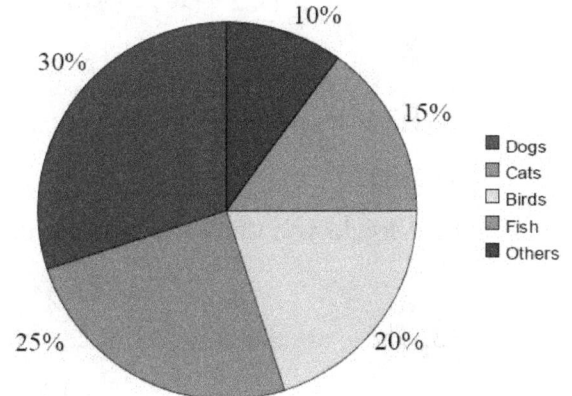

11. Teachers of a high school are asked about their favorite film genres. The answers were:

Comedy: 27%   Action: 18%

Romance: 14%   Drama: 14%

Horror: 11%   Foreign: 8%   Fiction: 8%

a) Draw a bar chart and a pie chart with the data.

b) Imagine there are 120 teachers in this high school. How many of them did answer "Romance"?

12. These are the number of photographic cameras sold the last year in a store:

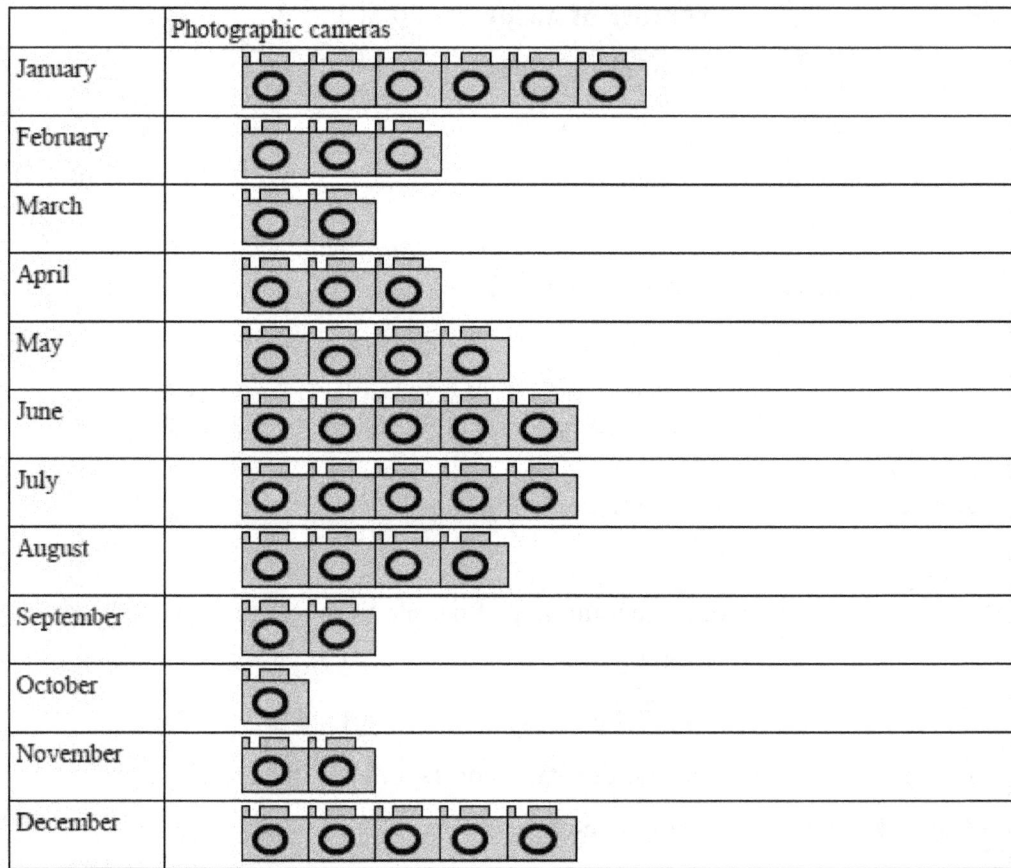

	Photographic cameras
January	⬤⬤⬤⬤⬤⬤
February	⬤⬤⬤
March	⬤⬤
April	⬤⬤⬤
May	⬤⬤⬤⬤
June	⬤⬤⬤⬤⬤
July	⬤⬤⬤⬤⬤
August	⬤⬤⬤⬤
September	⬤⬤
October	⬤
November	⬤⬤
December	⬤⬤⬤⬤⬤

⬤ = 10 photographic cameras

a) How many cameras has been sold every month? Which one was the best month for the salesman?

b) Why do you think was this month? What is the average of sold cameras per week?

13. These are the numbers of cars the different families of a street have:

1 2 1 1 0 2 3 1 1 2

3 1 0 2 1 2 1 1 2 1

a) Complete the table with the frequencies: f_i, h_i, F_i and H_i.

b) Represent the data, using a pie chart and a bar graph.

c) Find the mean, median and mode of the data.

14. Classify the following experiments into deterministic and random:

a) Extract a card from a pack of cards.

b) Measuring the hypotenuse of a right triangle whose legs are 3 cm and 4 cm.

c) Pushing "ON" in a lamp.

d) Measuring the height of a classroom.

e) Throwing a stone and measuring gravity constant.

f) Finding out the result of a football match before being played.

15. Write the sample set of the following random experiments:

a) Extract a card from a pack of Spanish cards.

b) Extracting a ball from a box containing 5 red, 3 blue and 2 green ones.

c) Throwing two dice and subtracting the results.

d) Throwing two dice and multiplying the results.

16. We throw a dodecahedron shaped die with faces numbered from 1 to 12.

a) What is the sample set?

b) Write the events: A = less than 5;   B = more than 4;

C = even number;   D = Not a multiple of 3.

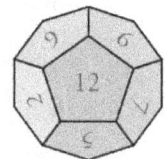

17. We write each letter of the word GAME in a different card, insert them into a bag and extract one.

a) Write the sample set.   b) Describe the event "obtaining a vowel".

c) If the word was PROBABILITY, how would you answer?

18. Consider the experiment in which we randomly get a domino token.

a) Describe the event A = Choosing a token whose numbers' sum is 6.

b) Describe the event B = Choosing a token whose numbers' product is 12.

19. In a box there are 15 numbered balls from 1 to 15 and we extract one of them. Write the elements of the following events:

a) Multiple of 3   b) Multiple of 2   c) Higher than 4

d) higher than 3 and lower than 8   e) Odd number.

20. Calculate the probability of obtaining a 5 when throwing each of the following dice:

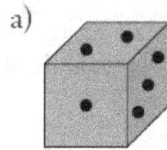

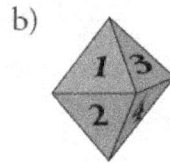

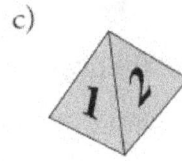

a)   b)   c)

21. In a bag there are 6 red bolls, 4 blue, 7 green, 2 yellow and 1 black one. We randomly extract one of them. Calculate the probability of obtaining one being:

   a) Blue   b) Not black c) Red or green   d) Nor yellow neither black.

22. In a test exam there are 80 units, from which you choose one randomly. A person has studied 60 of them. Calculate the probability of a) passing and b) failing the exam.

23. Calculate the probabilities of the following events when throwing a correct die:

   a) Multiple of 3   b) Multiple of 2   c) Higher than 1   d) Lower than 5
   e) Lower than 1

24. In a school, there are 990 students, being 510 of them, girls. If we randomly choose a student, what is the probability of choosing a boy?

25. In a high school, students are distributed as indicated in the table:

1st ESO	2nd ESO	3rd ESO	4th ESO	1st BACH	2nd BACH
210	250	260	220	140	120

   If we randomly choose a student, what is the probability that he/she studies:
   a) 3rd ESO   b) ESO   c) Bachiller

26. We have a card having a white and a black faces. We throw it 3 times. What is the probability of obtaining the black face a) exactly once?   b) twice and b) three times?

27. Probability of an event is 0.2. What is the probability of its complementary?

28. In a book having 120 pages, we have counted the number of misprints in each page. Results are in the following table. When randomly chosen a page of this book:

Number of errors	Number of pages
0	58
1	42
2	16
3	3
4	1

a) What is the probability of choosing a page without misprints?

b) What is the probability of choosing a page having exactly 2 misprints?

c) What is the probability of choosing a page having misprint/s? And having more than 3?

29. We randomly extract a domino token. Calculate the probability of getting:

a) A sum of punctuations being lower than 4.

b) A sum of punctuations being multiple of 3.

c) A "Double" token (having both identical punctuations).

30. We throw two dice: Calculate the probability of obtaining

a) A product of punctuations of 5.   b) A product of punctuations of 6.

c) A product of punctuations of 4.   *(Clue: build a table with possible results).*

31. We throw 100 times a tetrahedron shaped die (faces: 1, 2, 3 and 4), writing the number of the hidden face. We have obtained the following results:

Face	1	2	3	4
Frequency	28	22	30	20

Calculate the probabilities of the following events:

a) Multiple of 3   b) Multiple of 2   c) Higher than 1   d) Lower than 1.

32. Consider a game in which you throw two dice and you win if you obtain a sum of punctuations of 11 or 7. What is your probability of winning?

33. Numbers in a bingo's roulette are from 1 to 36. Calculate the following probabilities:

    a) Getting number 9.   b) Getting an even number.

    c) Getting a number finishing in a 3.

    d) Getting a multiple of 4

    e) Getting a number having two identical digits

    f) Getting a number starting with a 4

34. At home, throw a coin on the table for 10 times. Write down the number of times you obtain each side of the coin. What is the frequency of each event? Repeat this experiment throwing the coin 50 and 100 times. What are the frequencies now? What do you observe?

35. John suspects that he has been given a "treated" die. He has thrown it on a table and written down the face he has obtained. The results are given in the following table:

Face:	1	2	3	4	5	6
Number of times:	18	21	18	22	21	20

Calculate the frequency for each face. Do you think this die is correct or not?

www.ingramcontent.com/pod-product-compliance
Lightning Source LLC
Chambersburg PA
CBHW051336200326
41519CB00026B/7441